Introduction to
ABSTRACT ALGEBRA

Second Edition

THOMAS A. WHITELAW, B.Sc., Ph.D.

Lecturer in Mathematics
University of Glasgow

Blackie

Glasgow and Lo

D0807913

Blackie and Son Limited,
Bishopbriggs, Glasgow G64 2NZ
7 Leicester Place, London WC2H 7BP

British Library Cataloguing in Publication Data
Whitelaw, Thomas A.
Introduction to abstract algebra.——2nd ed.
1. Algebra, Abstract
I. Title
512'.02   QA162

ISBN 0-216-92259-3

Photosetting by Advanced Filmsetters (Glasgow) Ltd.
Printed in Great Britain by Bell & Bain (Glasgow) Ltd.

Preface

THIS IS A TEXTBOOK FOR STUDENTS UNDERTAKING, OR ABOUT TO UNDERTAKE, A first course in Abstract Algebra (e.g. second-year Honours students and second- or third-year Ordinary Degree students in Scottish universities; first-year students in many English universities). The first edition of the book proved helpful to many such students, and it is hoped that this second edition may be found even more helpful. The main changes made in producing the second edition have been the inclusion of a whole new chapter on the symmetric group and the expansion of the existing sets of exercises, together with updating and revising of the text.

The book is not an advanced treatise on group theory or on any kindred part of mathematics. But, in contrast to many elementary textbooks, it does penetrate far enough into Abstract Algebra to let the student see that the subject has worthwhile insights to offer, and to introduce him to some of the distinctive ways of thinking that produce interesting results. For example, in group theory, the material covered includes cyclic groups, Lagrange's theorem, homomorphisms, normal subgroups, quotient groups, and (in the new chapter) the partition of the symmetric group of degree n into conjugacy classes and an introduction to the alternating group of degree n. Although there is only one chapter on rings (and fields), it is a lengthy chapter and covers a wide variety of ideas.

Experience shows that students find Abstract Algebra difficult to grasp if they start with an inadequate understanding of basic topics (such as equivalence relations, mappings, and the notations and vocabulary of set theory.) Accordingly, as a preliminary to the chapters on groups, rings, etc., the first four chapters of this book deal with all the relevant details of those basic topics, along with certain properties of the integers. This last item is included to establish machinery (e.g. division algorithm) used in group theory and elsewhere, and also to provide material which will serve to motivate and illustrate some of the important concepts of Abstract Algebra. Chapter 1 also includes some material that comes under the heading *Logic*, and consists of explanations and advice about quantifiers, the use of the 'implies' sign, and proof by contradiction. The book demands little in the way of pre-requisites, though an elementary knowledge of complex numbers and matrices is assumed in some sections.

In writing this book, my aim throughout has been to make the exposition of ideas as clear as possible, *judged from the undergraduate student's point of view.* So, for example, in considering mappings, relations, etc., I have avoided presentations which, though satisfyingly neat and sophisticated to those who have long understood the concepts, do little to communicate them to the uninitiated.

The text includes worked examples and these have been chosen to reveal important methods of argument and show how certain kinds of proof may be set down. At the end of each chapter there is an ample supply of graded exercises, and partial solutions to these are at the back of the book. This adjunct to the exercises will, if sensibly used, prove beneficial to the majority of students.

I should like to record openly how much the merits of this book owe to the tradition of care and enthusiasm in the teaching of mathematics that has long existed in the Mathematics Department of the University of Glasgow. The traditions I refer to nourished me as an undergraduate; and, more recently, in working and conversing with my colleagues, I have learned much that has enriched my ideas about how mathematics can and should be taught. Special thanks go to my colleague Dr. J. B. Hickey for the help he has given during the preparation of this book. His kindness and his attention to detail have far exceeded all that I could have asked for.

T.A.W.

Contents

CHAPTER ONE

SETS AND LOGIC

1. Some very general remarks

This opening section is intended as an introduction to the entire book and not merely to the first chapter. It begins with an attempt to give brief answers to the questions: What is Abstract Algebra? and Why is it a worthwhile subject of study?

The thoughtful student may have realized that he has had to learn the properties of several different algebraic systems—the system of integers, the system of real numbers, the system of complex numbers, systems of matrices, systems of vectors, etc. He would be correct to guess that the number of algebraic systems he has to deal with will multiply rapidly as he proceeds further in the study of mathematics. This being the case, it is desirable to try to gain general insights about algebraic systems and to prove theorems that apply to broad classes of algebraic systems, instead of devoting energy to the study of a seemingly endless sequence of individual systems. If we can achieve such general insights, we stand to gain a deeper and more unified understanding of numerous elementary matters such as properties of integers or properties of matrices; and we shall be equipped to think intelligently about new algebraic systems when we encounter them. Thus there are ample reasons for believing it worthwhile (perhaps indeed essential) to attempt the study of algebraic systems in general, as opposed to the separate study of many individual algebraic systems. It is such study of algebraic systems in general that is called Abstract Algebra, and this is the part of mathematics to which this textbook provides an introduction.

The word *abstract* must not be taken to mean either difficult or unrelated to the more tangible objects of mathematical study (such as numbers and matrices). Instead, *abstract* indicates that the discussion will not be tied to one particular concrete algebraic system (since generality of insight is the paramount aim), while every general result obtained will have applications to a variety of different concrete situations.

1

In this book the study of abstract algebra proper begins at chapter five. Before that, attention is given to preliminary material of fundamental importance, including sets, equivalence relations, and mappings.

This first chapter contains a survey of the main ideas used continually in the discussion of sets. Interwoven with that are some remarks on elementary logical matters such as quantifiers, the implies sign, and proof by contradiction. It is hoped that these remarks will help students both to digest logical arguments that they read and to execute their own logical arguments. Even so, a clear warning must be given that mastery of the art of logical argument (as required in abstract algebra and comparable mathematical disciplines) depends less on knowledge of any core of facts about logic than on commitment to complete precision in the use of words and symbols.

2. Introductory remarks on sets

(a) Very often in mathematics we discuss sets or collections of objects and think of each set as a single thing that may be denoted by a single letter (S or A or ...).

(b) Standard notations for familiar sets of numbers are:

$\mathbb{N}$ denoting the set of all natural numbers (i.e. positive integers);
$\mathbb{Z}$ denoting the set of all integers;
$\mathbb{Q}$ denoting the set of all rational numbers;
$\mathbb{R}$ denoting the set of all real numbers;
$\mathbb{C}$ denoting the set of all complex numbers.

(c) Suppose that S is a set and x is an object. If x is one of the objects in the set S, we say that x is a **member** of S (or x is an **element** of S, or x **belongs to** S) and we indicate this by writing

$$x \in S.$$

In the contrary case, i.e. if x is not a member of S, we write

$$x \notin S.$$

We also use the $\in$ symbol as an abbreviation for *belonging to*, as in the phrase *for every $x \in S$* (read *for every x belonging to S*).

(d) The set whose members are all the objects appearing in the list $x_1, x_2, \ldots, x_n$ and no others is denoted by

$$\{x_1, x_2, \ldots, x_n\}.$$

(e) A set S is called **finite** if only finitely many different objects are members of S. Otherwise S is described as an infinite set.

If S is a finite set, the number of different objects in S is called the **order** (or **cardinality**) of S and denoted by $|S|$; e.g. if $S = \{1, 3, 5, 7\}$, then $|S| = 4$.

(f) It is important to have a clear criterion for deciding when two sets S_1, S_2 should be called *equal*. We adopt the natural definition that "$S_1 = S_2$" shall mean that every member of S_1 is a member of S_2 and every member of S_2 is a member of S_1.

(g) It is already apparent that we are basing our discussion of sets on the simplistic assumption that we can recognize a "set" or "collection of objects" when we meet one. In fact, some far-reaching objections can be raised to this approach. Most seriously, it can be shown that contradictions arise when we apply our simple-minded notions about sets to collections of objects so vast that we cannot really discern whether they fit in with the simple-minded notions. The circumvention of these difficulties lies beyond the scope of this book. Here our attitude is to assume provisionally that the simple-minded concept of a set is a reasonable and useful one in the situations we are exploring, and to concentrate on explaining facts and notations that are frequently used in these situations. (Students who would like to know more about the problems we have hinted at should read the earlier chapters of *Naive Set Theory* by P. R. Halmos [1].)

3. Statements and conditions; quantifiers

In thinking generally about the sentences that occur in mathematics, it is valuable to draw a distinction between statements and conditions.

A *statement* means a sentence about specific numbers, sets, or other objects, the sentence being, in its own right, either true or false (even if it is difficult to say which). So examples of statements include:

$$3 \times 4 = 11$$

$$2^{19937} - 1 \text{ is a prime number.}$$

A *condition*, on the other hand, is a sentence (such as "$x^3 > 3x$" or "$y^2 = 4x$") that involves one or more variables (symbols like $x, y, \ldots$) and becomes a statement when each variable is given a specific value (i.e. is replaced by a definite object). There should always be an indication of the set of values each variable can take (i.e. be replaced by); and for each variable this set is called the **domain** of the variable.

For illustrative purposes, consider for a moment one particular example of a condition, namely

$$y^2 = 4x \qquad (x, y \in \mathbb{R}).$$

The bracketed "$x, y \in \mathbb{R}$" indicates that both x and y have $\mathbb{R}$ as domain. From this condition, true statements can be produced by giving x, y certain specific values (e.g. giving x the value 9 and y the value 6 produces the true statement $6^2 = 4 \times 9$); and false statements can be produced by giving x, y other values.

There is a wide variety of accepted ways of expressing the fact that a true statement is produced from the condition on replacing x by 9 and y by 6: for example, one may say that "the condition is true for $x = 9$, $y = 6$" or "the condition holds when $x = 9$, $y = 6$" or "the condition is satisfied by $x = 9$, $y = 6$", etc.

The commonest kind of condition is the one-variable condition, i.e. condition involving just one variable. Such a condition may conveniently be denoted by $P(x)$, x being the variable with, let us suppose, domain S. We denote by

$$\{x \in S : P(x)\}$$

the set of elements in S that satisfy the condition $P(x)$, i.e. the set of values of x in S for which $P(x)$ is true. For example,

$$\{x \in \mathbb{Z} : x^2 = 4\} = \{-2, 2\}.$$

The universal quantifier

Suppose here that $P(x)$ is a condition involving the variable x, which has domain S.

The symbol $\forall$ is shorthand for "for all" or "for every" and may be used to build the statement

$$\forall x \in S,\ P(x), \tag{1}$$

which is the assertion that $P(x)$ is true for all values of x in the set S. It should be noticed that this new sentence is not a condition but a statement, true or false in its own right—true if $P(x)$ is true for all values of x in S, false otherwise. A statement of the form (1) is called a **universal statement**, and the symbol $\forall$ is called the **universal quantifier.**

As specific illustrations of universal statements, consider:

$$\forall x \in \mathbb{R},\quad (x+1)^2 = x^2 + 2x + 1,$$
$$\forall x \in \mathbb{R},\quad (x+1)^2 = x^2 + 1.$$

Elementary algebra tells us that the former is true. But the latter is false, since the condition "$(x+1)^2 = x^2 + 1$" is false for certain values of x (e.g. $x = 1$).

Remarks. (*a*) The assertion "$\forall x \in S, P(x)$" is equivalent to the assertion that the set $\{x \in S : P(x)\}$ is the whole of S.

(*b*) Universal statements can also be made from conditions involving more than one variable. If $P(x, y, z, \ldots)$ is a condition involving the variables $x, y, z, \ldots$, with respective domains $A, B, C, \ldots$, then the statement

$$\forall x \in A,\quad \forall y \in B,\quad \forall z \in C, \ldots,\quad P(x, y, z, \ldots)$$

is the assertion that $P(x, y, z, \ldots)$ is true for all relevant values of $x, y, z, \ldots$ (i.e. all values of x in A, etc.). For instance, a familiar index law is:

$$\forall x \in \mathbb{R}, \quad \forall m \in \mathbb{N}, \quad \forall n \in \mathbb{N}, \quad x^m . x^n = x^{m+n}.$$

(c) It should be observed that all of the following are synonymous:

$$\forall x \in \mathbb{R}, \, x^2 . x^3 = x^5; \quad \forall y \in \mathbb{R}, \, y^2 . y^3 = y^5; \quad \forall \alpha \in \mathbb{R}, \, \alpha^2 . \alpha^3 = \alpha^5; \quad \text{etc.}$$

Each of them says, in formal language, that the product of the square of a real number and its cube is always equal to the fifth power of that number. More generally, the fact is that, in a universal statement, one can change the symbol used for the variable quite arbitrarily without altering the meaning of the statement. This is often expressed by saying that the variable is a *dummy*.

(d) A well-established convention in mathematicians' use of language is that where an asserted fact takes the form of a condition (accompanied by an indication of the domain(s) of the variable(s)), it is to be understood that what is being asserted is the truth of the condition for all values of the variables. For example, one may find asserted as a theorem

$$x^m . x^n = x^{m+n} \qquad (x \in \mathbb{R}, \, m, n \in \mathbb{N}).$$

This must be interpreted as meaning

$$\forall x \in \mathbb{R}, \quad \forall m \in \mathbb{N}, \quad \forall n \in \mathbb{N}, \quad x^m . x^n = x^{m+n}.$$

The existential quantifier
Suppose again that $P(x)$ is a condition involving the variable x, which has domain S.

The symbol $\exists$ is shorthand for "there exists" and may be used to build the statement

$$\exists x \in S \text{ s.t. } P(x)$$

(read "there exists x belonging to S such that $P(x)$"), which is the assertion that there is at least one value of x (in S) for which $P(x)$ is true.

Thus, for instance,

$$\exists x \in \mathbb{R} \text{ s.t. } x^2 = 4$$

is true: for the condition $x^2 = 4$ is satisfied by at least one value (in $\mathbb{R}$) of x (e.g. $x = 2$). But

$$\exists x \in \mathbb{R} \text{ s.t. } x^2 = -1$$

is false: for there is no real value of x satisfying $x^2 = -1$.

It should be noted that a sentence of the form "$\exists x \in S \text{ s.t. } P(x)$" is a statement, not a condition, as it is true or false in its own right. Such a statement is called an **existential statement**, and the symbol ∃ is called the **existential quantifier.**

Remarks. (*a*) The assertion "$\exists x \in S$ s.t. $P(x)$" is equivalent to the assertion that the set $\{x \in S : P(x)\}$ has at least one member.

(*b*) It should be observed that the following are synonymous:

$$\exists x \in \mathbb{R} \text{ s.t. } x^2 = 4; \quad \exists y \in \mathbb{R} \text{ s.t. } y^2 = 4; \quad \exists \alpha \in \mathbb{R} \text{ s.t. } \alpha^2 = 4; \quad \text{etc.}$$

Once again, we describe this situation by saying that the variables are dummies.

Negation of a statement

For any given statement p, the **negation** of p is the statement "not p" (denoted $\neg p$) produced from p by inserting the word "not" at the appropriate point so as to produce a statement denying precisely that which p asserts. For example, if p is "$2^{100} > 3^{63}$", then $\neg p$ is "2^{100} is not greater than 3^{63}" (or, synonymously, $2^{100} \leqslant 3^{63}$). Obviously $\neg(\neg p)$ is always synonymous with p.

Likewise, from any condition $P(x)$ we can produce its negation, $\neg P(x)$, by inserting "not" at the crucial point, so that if $P(x)$ is true for a certain value of x, then $\neg P(x)$ is false for that value of x, and vice versa.

The following general rules tell us that the negation of a universal statement can be re-expressed as an existential statement, and vice versa.

3.1 (i) $\neg[\forall x \in S, P(x)]$ is synonymous with $\exists x \in S$ s.t. $\neg P(x)$.

(ii) $\neg[\exists x \in S \text{ s.t. } P(x)]$ is synonymous with $\forall x \in S, \neg P(x)$.

Proof. (i) Let α stand for "$\neg[\forall x \in S, P(x)]$" and β stand for "$\exists x \in S$ s.t. $\neg P(x)$".

The crux of the matter is that "not everyone does" says the same as "at least one doesn't". Therefore α, which asserts that not every value of x (in S) satisfies $P(x)$, is equivalent to the assertion that at least one value of x (in S) does not satisfy $P(x)$, i.e. that at least one value of x (in S) satisfies $\neg P(x)$: and this is precisely what β asserts.

(ii) can be justified similarly by observing that "no one does" says the same as "everyone does not".

To give a very simple illustration of 3.1, let us suppose that S is a class of students and that $P(x)$ stands for "student x is aged 21 or over" (so that $\neg P(x)$ means "student x is aged under 21"). Applied in this case, 3.1(i) tells us that the negation of "every student in the class is aged 21 or over" is "there exists a student in the class aged under 21"; and 3.1(ii) tells us that the negation of "there exists a student in the class aged 21 or over" is "every student in the class is aged under 21".

These deductions from 3.1 may rightly be considered to be simple common sense: the point of 3.1 is that it brings out the general principle vaguely discerned in such "common sense" and equips us to apply the principle in more complicated contexts.

Proofs of universal and existential statements
Notice first that to prove that an existential statement is true, we need only consider one (appropriately chosen) value of the variable. For example, to prove true "$\exists x \in \mathbb{R}$ s.t. $x^2 = 4$", we need only produce one value of x (in $\mathbb{R}$) satisfying $x^2 = 4$; and thus it suffices to consider $x = 2$.

An existential statement being the negation of a universal statement, it is also evident that a universal statement may be proved false by citing just one (appropriately chosen) value of the variable. For example, to prove that "$\forall x \in \mathbb{R}, (x+1)^2 = x^2 + 1$" is false, we need only exhibit one value of x (in $\mathbb{R}$) that does not satisfy the condition $(x+1)^2 = x^2 + 1$ (and clearly 1 is such a value of x).

A value of x for which $P(x)$ is false, exhibited to disprove the universal statement "$\forall x \in S, P(x)$", is called a **counterexample** to this universal statement.

Obviously, however, one cannot hope to prove the truth of a true universal statement merely by considering a single value of the variable! The basic ideas generally used for tackling and setting down proof of a universal statement are perhaps most easily absorbed by study of a very straightforward illustration. For this purpose a proof will now be given of the universal statement

$$\forall x \in E, \quad x+4 \text{ is even}, \tag{2}$$

where E denotes the set of all even integers. [Here and elsewhere in this chapter, it must be explained, we shall work from the following assumptions about even and odd integers: (i) an even integer is one expressible as $2 \times$ (an integer); (ii) an odd integer is one expressible as $2 \times$ (an integer) $+ 1$; (iii) every integer is either even or odd (and not both).]

Proof of (2). Let x be an arbitrary member of E.
Then $x = 2k$ for some integer k, and hence

$$x+4 = 2k+4 = 2(k+2).$$

Therefore, since $k+2$ is an integer, $x+4$ is even.
It follows that $\forall x \in E, x+4$ is even.

In the above proof, note particularly the device of opening by saying, "Let x be an arbitrary member of ...". Some may be unhappy about a more pedantic point—the double use of the symbol x (as dummy variable in the universal statement and as arbitrary element of a set in the proof); but an assurance can be given that this very widespread habit does not in practice cause confusion.

To conclude the section, notice that disproof of an existential statement, which has not been separately discussed, is (by 3.1(ii)) equivalent to proof of a universal statement.

4. The implies sign ($\Rightarrow$)

Here let P, Q stand either for statements or for conditions involving the same variables $x, y, \ldots$. The compound sentence

$$P \Rightarrow Q$$

(read "P implies Q" or, alternatively, "if P then Q") is the assertion that in every admissible case when P is true, Q is true too, i.e. that in no allowed case do we have P true and Q false. For example,

$$x > 2 \Rightarrow x^2 > 4 \qquad (x \in \mathbb{R})$$

is true because every value of x (in $\mathbb{R}$) that satisfies the condition "$x > 2$" satisfies also the condition "$x^2 > 4$". But

$$x^2 > 4 \Rightarrow x > 2 \qquad (x \in \mathbb{R})$$

is false, as we can show by exhibiting a value of x (in $\mathbb{R}$) that satisfies "$x^2 > 4$" but does not satisfy "$x > 2$" (and $x = -3$ is such a counterexample).

Remarks. (*a*) If $P(x)$, $Q(x)$ are conditions involving the variable x with domain S, the sentence

$$P(x) \Rightarrow Q(x) \qquad (x \in S) \tag{3}$$

is really a universal statement, because it says that a certain condition is satisfied for every value (in S) that might be given to x (namely that the combination $P(x)$ true, $Q(x)$ false does not arise). Purists might therefore prefer that (3) should be re-written

$$\forall x \in S, \quad P(x) \Rightarrow Q(x)$$

though in fact (3) is typical of the way mathematicians usually express themselves.

Notice also that the universal statement "$\forall x \in S, P(x)$" can be re-written as

$$x \in S \Rightarrow P(x) \qquad (x \in S).$$

(*b*) "$P \Rightarrow Q$" is held to be true (by default, as it were) if there is no case in which P is true; e.g. the statement

$$x \neq x \Rightarrow x = 7 \qquad (x \in \mathbb{R})$$

is classified as true, because in no allowed case do we have $x \neq x$ true and $x = 7$ false.

(c) It is essential to realize that the statement "$P \Rightarrow Q$" says something about the *relationship between P and Q*. It does not say anything about the truth of P in any or all cases; nor does it say anything about the truth of Q in any or all cases. In particular "$\Rightarrow$", which comes in the middle of a compound sentence, is most certainly not a synonym for "hence" or "therefore", which are appropriate words at the beginning of a new sentence. If, in the course of a proof, we wish to assert that, because of what has just been said, the truth of Q is apparent, then we should say "Hence Q" or "Therefore Q" or "It follows that Q" (but not "$\Rightarrow Q$").

Formats for the proof of "$P \Rightarrow Q$"
A straightforward way of giving such a proof is to begin by saying "Suppose P" or "Assume P", and then to proceed to demonstrate that Q is a consequence of this supposition. Thus, by temporarily regarding P as true (for the purpose of the argument), we show that we cannot have P true without also having Q true—which is precisely what "$P \Rightarrow Q$" means.

As an illustration, here is a proof of the simple result

$$n \text{ is odd} \Rightarrow n^2 \text{ is odd} \qquad (n \in \mathbb{Z}).$$

Proof. Suppose that n is odd ($n \in \mathbb{Z}$). Then $n = 2k+1$ for some $k \in \mathbb{Z}$, and hence

$$n^2 = (2k+1)^2 = 4k^2 + 4k + 1 = 2(2k^2 + 2k) + 1.$$

Therefore, since $2k^2 + 2k \in \mathbb{Z}$, n^2 is odd.
The stated result follows.

Another common form of proof of "$P \Rightarrow Q$" is a sequence

$$P \Rightarrow \alpha_1 \Rightarrow \alpha_2 \Rightarrow \alpha_3 \Rightarrow \ldots \Rightarrow \alpha_n \Rightarrow Q. \qquad (4)$$

In interpreting such a proof, it is essential to appreciate that (4) is an abbreviation for:

$$\begin{cases} P \Rightarrow \alpha_1, \alpha_1 \Rightarrow \alpha_2, \alpha_2 \Rightarrow \alpha_3, \ldots, \alpha_n \Rightarrow Q \text{ are all true.} \\ \text{Consequently, } P \Rightarrow Q \text{ is true.} \end{cases}$$

The logic underlying the word "consequently" is not hard to discern.

As a simple illustration of the format (4), here is a proof of this type of the result

$$n \text{ is odd} \Rightarrow n^2 \text{ is odd} \qquad (n \in \mathbb{Z}).$$

Proof. For $n \in \mathbb{Z}$,

$$n \text{ is odd} \Rightarrow n = 2k+1 \ (k \text{ being an integer})$$
$$\Rightarrow n^2 = (2k+1)^2 = 2(2k^2+2k)+1$$
$$\Rightarrow n^2 \text{ is odd (since } 2k^2+2k \text{ is an integer).}$$

The law of contraposition

4.1 "$P \Rightarrow Q$" is synonymous with "$(\neg Q) \Rightarrow (\neg P)$".

Proof. "$P \Rightarrow Q$" says that it is not possible to have P true and Q false.

"$(\neg Q) \Rightarrow (\neg P)$" says that it is not possible to have $\neg Q$ true and $\neg P$ false, i.e. to have Q false and P true.

Thus the two are synonymous.

We call "$(\neg Q) \Rightarrow (\neg P)$" the **contrapositive** of "$P \Rightarrow Q$", and we call the very useful law 4.1 (the fact that "$P \Rightarrow Q$" is synonymous with its contrapositive) the *law of contraposition*.

In contrast, the sentence "$Q \Rightarrow P$" is called the **converse** of "$P \Rightarrow Q$". A common and bad error is to confuse "$P \Rightarrow Q$" and its converse, or to believe that they are synonymous. That, let it be emphasized, is certainly not so. Indeed we observed early in the section that, while

$$x > 2 \Rightarrow x^2 > 4 \qquad (x \in \mathbb{R})$$

is true, its converse is false.

"If and only if" ($\Leftrightarrow$)

The compound sentence "$P \Leftrightarrow Q$" (read "P if and only if Q" or "P is equivalent to Q") means "both $P \Rightarrow Q$ and $Q \Rightarrow P$". Thus, when we assert the truth of "$P \Leftrightarrow Q$", we are asserting the truth both of "$P \Rightarrow Q$" and of its converse "$Q \Rightarrow P$", and we are therefore claiming that, in every admissible case, P and Q are either both true or both false.

Sometimes, especially in definitions, it is convenient to employ the abbreviation "iff" (rather than the symbol $\Leftrightarrow$) for the words "if and only if".

The proof of "$P \Leftrightarrow Q$" must often be split into two parts—the first a proof of "$P \Rightarrow Q$" and the second a proof of "$Q \Rightarrow P$". Sometimes we encounter a proof of "$P \Leftrightarrow Q$" taking the form of a sequence

$$P \Leftrightarrow \alpha_1 \Leftrightarrow \alpha_2 \Leftrightarrow \alpha_3 \Leftrightarrow \ldots \Leftrightarrow \alpha_n \Leftrightarrow Q.$$

This (cf. corresponding remarks on proving "$P \Rightarrow Q$") must be understood as an abbreviation for:

$$\left\{ \begin{array}{l} P \Leftrightarrow \alpha_1, \alpha_1 \Leftrightarrow \alpha_2, \alpha_2 \Leftrightarrow \alpha_3, \ldots, \alpha_n \Leftrightarrow Q \text{ are all true.} \\ \text{Consequently, } P \Leftrightarrow Q \text{ is true.} \end{array} \right\}$$

5. Proof by contradiction

In many contexts a powerful method of proof is *proof by contradiction*, which, in effect, consists of eliminating the possibility that the statement to be proved is false. Those unfamiliar with the method are recommended to study the next two paragraphs of general explanation in conjunction with the simple example which follows them.

When using proof by contradiction, we proceed as follows.

(*a*) We begin by assuming, for the sake of argument, that the statement to be proved is false.

(*b*) We deduce from that assumption a contradiction, i.e. something that is manifestly false because it is self-contradictory (e.g. "n is both zero and nonzero") or contrary to known facts (e.g. "$2 = 1$") or contrary to what was given or assumed at stage (*a*).

(*c*) We infer that the assumption made at stage (*a*) was wrong—which is to say that the statement to be proved is in fact true.

The thinking behind the vital step (*c*) is that the obtaining through accurate logical deduction of a contradiction reveals that at some point we have made a wrong assumption (since valid assumptions lead only to true conclusions): so, since the only questionable assumption made was the one in stage (*a*), that assumption must be wrong.

Example. It is given that m, n are integers and that the product mn is odd. Prove that m is odd.

Proof. Suppose the contrary, i.e. that m is even. Then $m = 2k$ for some integer k, and hence

$$mn = 2kn = 2 \times (\text{an integer}).$$

So mn is even—a contradiction [i.e. something contrary to what is given].

From this contradiction it follows that m is odd.

6. Subsets

Let A, B denote sets. We say that A is a **subset** of B (or that A is **contained in** B) iff every member of A is also a member of B.

In this book, our notation for "A is a subset of B" will be

$$A \subseteq B.$$

When $A \subseteq B$ and $A \neq B$, we say that A is a **proper subset** of B (or that A is **strictly contained** in B) and we write $A \subset B$. [The student is advised that mathematicians (and the textbooks they write) vary slightly in their usages of the symbols $\subseteq$ and $\subset$.]

Clearly:

6.1 For every set A, $A \subseteq A$.
6.2 For sets A, B, C, $(A \subseteq B$ and $B \subseteq C) \Rightarrow A \subseteq C$.
6.3 For sets A, B, $A = B \Leftrightarrow (A \subseteq B$ and $B \subseteq A)$.

Remarks. (a) If A, B are subsets of the set S, the statement "$A \subseteq B$" is equivalent to

$$x \in A \Rightarrow x \in B \qquad (x \in S),$$

and the statement "$A = B$" is equivalent to

$$x \in A \Leftrightarrow x \in B \qquad (x \in S).$$

These simple observations are the basis of many proofs of statements of the form "$A \subseteq B$" or "$A = B$".

(b) The notation $\{x \in S : P(x)\}$ for specifying a subset of the set S was introduced in §3. Notice that in this notation x is a dummy variable. For example,

$$\{x \in \mathbb{R} : x^2 = 4\}, \quad \{y \in \mathbb{R} : y^2 = 4\}, \quad \{\alpha \in \mathbb{R} : \alpha^2 = 4\}, \quad \text{etc.}$$

all mean the same thing, viz. $\{-2, 2\}$.

(c) In illustrations we shall often refer to **intervals**, which are "connected" subsets of $\mathbb{R}$. Specifically, if $a, b \in \mathbb{R}$ and $a < b$, the intervals with a and b as endpoints are the four subsets

$$(a, b) = \{x \in \mathbb{R} : a < x < b\}, \qquad [a, b] = \{x \in \mathbb{R} : a \leqslant x \leqslant b\},$$
$$(a, b] = \{x \in \mathbb{R} : a < x \leqslant b\}, \qquad [a, b) = \{x \in \mathbb{R} : a \leqslant x < b\};$$

and, for each $a \in \mathbb{R}$, we write

$$(-\infty, a) = \{x \in \mathbb{R} : x < a\}, \qquad (-\infty, a] = \{x \in \mathbb{R} : x \leqslant a\},$$
$$(a, \infty) = \{x \in \mathbb{R} : x > a\}, \qquad [a, \infty) = \{x \in \mathbb{R} : x \geqslant a\}.$$

(d) Another notation used in specifying a subset of a set S takes the form

$$\{f(x) : x \in A\}$$

where A is a given set and $f(x)$ stands for something constructible from x (e.g. x^2). It is to be understood that if x is given a definite value in A, then correspondingly $f(x)$ will take a definite value in S. And $\{f(x) : x \in A\}$ means the set of all values $f(x)$ can take if x takes a value in A. For example,

$$\{x^2 : x \in \mathbb{N}\} = \{1, 4, 9, 16, 25, \ldots\};$$

and, if $A = \{1, -1, 2, -2, 3, -3\}$, then $\left\{\dfrac{1}{x^2 + 1} : x \in A\right\} = \{\tfrac{1}{2}, \tfrac{1}{5}, \tfrac{1}{10}\}$. In formal terms, $\{f(x) : x \in A\}$ can be defined as

$$\{y \in S : \exists x \in A \text{ s.t. } y = f(x)\}.$$

More complicated variants of this notation abound but are to some extent self-explanatory. When one occurs in the pages of this book, there will always be accompanying words aimed at making the meaning fully clear.

The empty set
An **empty set** is a set with no members: e.g. $\{x \in \mathbb{R} : x^2 = -1\}$.

If θ is an empty set and A is any set whatever, then it is true by default that every member of θ is a member of A, i.e. that

$$x \in \theta \Rightarrow x \in A \qquad (x \in A).$$

Hence:

6.4 If θ is an empty set and A is any set whatever, then $\theta \subseteq A$.

From this it can immediately be deduced that:
6.5 There is only one empty set.

Proof. If θ_1 and θ_2 are empty sets, then (by 6.4) $\theta_1 \subseteq \theta_2$ and $\theta_2 \subseteq \theta_1$, and hence (see 6.3) $\theta_1 = \theta_2$. The result follows.

Because of 6.5, it makes sense to speak of *the* empty set, which we always denote by θ.

Complement of a subset
Definition. Let A be a subset of a given set S. The **complement** of A in S is the subset $\{x \in S : x \notin A\}$. It is denoted by $\mathscr{C}_S A$.

For example, if E is the set of all even integers, then $\mathscr{C}_{\mathbb{Z}} E$ is the set of all odd integers.

When it is understood that a particular set S is our "universal set" (i.e. that all the sets being considered are subsets of S and that "complement of A" means "complement of A in S"), it is feasible to use simpler notations for $\mathscr{C}_S A$, e.g. $\mathscr{C} A$ or A'.

Clearly, when S is the universal set:

6.6 $\mathscr{C}\theta = S$; $\mathscr{C}S = \theta$; and, for every $A \subseteq S$, $\mathscr{C}(\mathscr{C}A) = A$.

Another definition. If A, B are sets, the **relative complement** of B in A (denoted by $A - B$) is the set $\{x \in A : x \notin B\}$. [Here, notice, we are not insisting that $B \subseteq A$. Obviously, when $B \subseteq A$, $A - B$ coincides with $\mathscr{C}_A B$.]

For example, if $A = \{1, 2, 3\}$ and $B = \{2, 4, 5, 6\}$, then $A - B = \{1, 3\}$.

Venn diagrams
When subsets of a set S are discussed, it is often helpful to draw pictures in which S is represented by a rectangular region (the elements of S being

represented by points of the region) and subsets $A, B, \ldots$ of S are represented by subregions of the rectangle. Such a pictorial representation is called a Venn diagram.

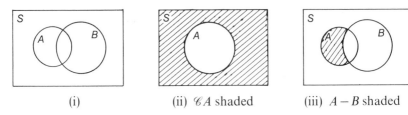

(i)   (ii) $\mathscr{C}A$ shaded   (iii) $A - B$ shaded

Parts (ii) and (iii) are Venn diagrams which, in an obvious way, illustrate how $\mathscr{C}A$ is related to A and how $A - B$ is related to A and B.

To be useful, a Venn diagram must represent the most general situation under discussion. Thus, for example, for considering two arbitrary subsets A, B of the set S, the Venn diagram (ii) below is a satisfactory visual aid while (i) is not, as it suggests that A, B have no members in common—which of course need not be the case.

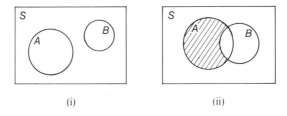

(i)   (ii)

Typical of the sort of proposition whose truth can, so to speak, be seen from a Venn diagram is:

6.7 For subsets A, B of a set S, $A \subseteq B \Leftrightarrow \mathscr{C}B \subseteq \mathscr{C}A$.

Justification by Venn diagram. We refer to diagram (ii). Both $A \subseteq B$ and $\mathscr{C}B \subseteq \mathscr{C}A$ are equivalent to the assertion that the set represented by the shaded region (i.e. $A - B$) is empty.

Formal proof. $A \subseteq B$ is equivalent to

$$x \in A \Rightarrow x \in B \qquad (x \in S).$$

By the law of contraposition (4.1), this is equivalent to

$$x \notin B \Rightarrow x \notin A \qquad (x \in S),$$

i.e.

$$x \in \mathscr{C}B \Rightarrow x \in \mathscr{C}A \qquad (x \in S),$$

and that is precisely what $\mathscr{C}B \subseteq \mathscr{C}A$ says.

Power set of a given set

Definition. For a given set S, the **power set** of S (denoted by $\mathscr{P}(S)$) is the set whose members are the subsets of S. (So $A \in \mathscr{P}(S)$ means $A \subseteq S$.)

For example, if $S = \{a, b\}$ (where $a \neq b$), then $\mathscr{P}(S)$ has 4 members, viz. $\emptyset$, S, $\{a\}$, $\{b\}$; i.e. $\mathscr{P}(S) = \{\emptyset, S, \{a\}, \{b\}\}$.

6.8 Let S be a finite set of order n. Then $|\mathscr{P}(S)| = 2^n$.

Sketch of proof. In choosing a subset of S, we have, for each of the n elements of S, 2 possibilities—inclusion in the subset or exclusion from it. Hence, in the choosing of a subset of S, the total number of possibilities is $2 \times 2 \times 2 \times \ldots$ to n factors, i.e. 2^n.

7. Unions and intersections

Consider first a sequence of n sets $A_1, A_2, \ldots, A_n$ (where $n \geqslant 2$). For clarity, we suppose that all of $A_1, A_2, \ldots, A_n$ are subsets of a universal set S, but we do not insist that the sets $A_1, A_2, \ldots, A_n$ must all be different from one another.

Definitions. (1) The **intersection** of the sets $A_1, A_2, \ldots, A_n$ is the subset of S comprising those elements that belong to all of $A_1, A_2, \ldots, A_n$. It is denoted by

$$A_1 \cap A_2 \cap \ldots \cap A_n \quad \text{or} \quad \bigcap_{r=1}^{n} A_r.$$

(2) The **union** of the sets $A_1, A_2, \ldots, A_n$ is the subset of S comprising those elements that belong to at least one of the sets $A_1, A_2, \ldots, A_n$. It is denoted by

$$A_1 \cup A_2 \cup \ldots \cup A_n \quad \text{or} \quad \bigcup_{r=1}^{n} A_r.$$

As an illustration, consider the following subsets of $\mathbb{N}$:

$$A_1 = \{1, 2, 3, 4\}, \quad A_2 = \{1, 2, 5\}, \quad A_3 = \{2, 4, 6, 8, 12\}.$$

In this case, as the student should verify to his satisfaction,

$$A_1 \cap A_2 \cap A_3 = \{2\}, \quad \text{and} \quad A_1 \cup A_2 \cup A_3 = \{1, 2, 3, 4, 5, 6, 8, 12\}.$$

It should be observed that a more formal phrasing of the above definitions brings in the quantifiers:

$$\bigcap_{r=1}^{n} A_r = \{x \in S : \forall r \in \{1, 2, \ldots, n\}, x \in A_r\},$$

$$\bigcup_{r=1}^{n} A_r = \{x \in S : \exists r \in \{1, 2, \ldots, n\} \text{ s.t. } x \in A_r\}.$$

The simplest case is the intersection or union of two sets A, B ($\subseteq S$), illustrated by the Venn diagrams below. Notice that

$$A \cap B = \{x \in S : x \in A \text{ and } x \in B\}$$
$$A \cup B = \{x \in S : x \in A \text{ or } x \in B\}$$

where, in the second line (as always in mathematics), we use the word "or" in the *inclusive* sense, i.e. we hold "P or Q" to be true whenever at least one of P, Q is true.

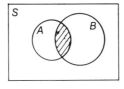 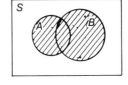

(i) $A \cap B$ shaded   (ii) $A \cup B$ shaded

Sometimes one has to deal with infinite families of sets. For example, one might have an infinite sequence of sets $A_1, A_2, A_3, \ldots$, or one might have a set A_x corresponding to each real number x. In all such cases the intersection and union of the family of sets are defined in the natural way: the intersection is the set consisting of the objects that belong to all of the sets in the given family; the union is the set consisting of the objects that belong to at least one of the sets in the given family.

Two further definitions. (1) Sets A, B are called **disjoint** iff $A \cap B = \emptyset$.

(2) If $\mathcal{T}$ is a collection of subsets of the set S (i.e. $\mathcal{T} \subseteq \mathcal{P}(S)$), we call $\mathcal{T}$ a **partition** of S iff the following conditions are all satisfied:

(i) the union of all the subsets belonging to $\mathcal{T}$ is the whole of S;

(ii) unequal members of $\mathcal{T}$ are disjoint;

(iii) each subset belonging to $\mathcal{T}$ is non-empty.

(Note that conditions (i) and (ii) are equivalent to the single condition that every element of S belongs to precisely one of the subsets in the collection $\mathcal{T}$.)

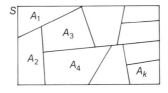

The Venn diagram illustrates the situation when $A_1, A_2, \ldots, A_k$ are different subsets forming a partition of S (i.e. the collection $\{A_1, A_2, \ldots, A_k\}$ is a partition of S).

There exist many interesting results about complements, unions and intersections of subsets, but to devote much space to them would be at variance with the basic purpose of this book. Accordingly, we shall here discuss only a few of these results, which have widespread uses or have proofs which instructively illustrate logical points described earlier in the chapter.

We consider first the *de Morgan laws* (7.1 below), which we state here for a finite family of sets, though the perceptive student will realize that, by the same approach, they can be proved true for a perfectly arbitrary family of sets.

7.1 For sets $A_1, A_2, \ldots, A_n$ (all subsets of universal set S),

$$\text{(i)} \quad \mathscr{C}\left(\bigcap_{r=1}^{n} A_r\right) = \bigcup_{r=1}^{n} (\mathscr{C}A_r),$$

$$\text{(ii)} \quad \mathscr{C}\left(\bigcup_{r=1}^{n} A_r\right) = \bigcap_{r=1}^{n} (\mathscr{C}A_r).$$

Proof. (i) We show (cf. remark (a) in §6) that, for $x \in S$,

$$x \in \text{left-hand side} \Leftrightarrow x \in \text{right-hand side}.$$

$$x \in \mathscr{C}\left(\bigcap_{r=1}^{n} A_r\right) \Leftrightarrow x \notin \bigcap_{r=1}^{n} A_r$$

$$\Leftrightarrow \neg\left[x \in \bigcap_{r=1}^{n} A_r\right]$$

$$\Leftrightarrow \neg\left[\forall r \in \{1, 2, \ldots, n\}, x \in A_r\right]$$

$$\Leftrightarrow \exists r \in \{1, 2, \ldots, n\} \text{ s.t. } \neg (x \in A_r) \qquad \text{(by 3.1(i))}$$

$$\Leftrightarrow \exists r \in \{1, 2, \ldots, n\} \text{ s.t. } x \in \mathscr{C}A_r$$

$$\Leftrightarrow x \in \bigcup_{r=1}^{n} \mathscr{C}A_r.$$

This establishes part (i).

Part (ii) may be proved similarly (using 3.1(ii)); or it may be deduced from (i) as follows.

Let $B_r = \mathscr{C}A_r$ $(r = 1, 2, \ldots, n)$. Then, by (i),

$$\mathscr{C}\left(\bigcap_{r=1}^{n} B_r\right) = \bigcup_{r=1}^{n} \mathscr{C}B_r, \quad \text{i.e.} \quad \mathscr{C}\left(\bigcap_{r=1}^{n} B_r\right) = \bigcup_{r=1}^{n} A_r.$$

The result now follows on taking the complement of each side.

It is clear (as the Venn diagrams on page 16 illustrate) that, for every pair of sets A, B, $A \cap B$ is a subset of A and of B, while both A and B are subsets of $A \cup B$. We now take note of a sharper result in this vein.

7.2 Let A, B be sets (subsets of universal set S). Then:

(i) $A \cap B$ is the largest subset of S contained in both A and B, in the sense that, for $C \subseteq S$, $(C \subseteq A$ and $C \subseteq B) \Rightarrow C \subseteq A \cap B$;

(ii) $A \cup B$ is the smallest subset of S containing both A and B, in the sense that, for $C \subseteq S$, $(A \subseteq C$ and $B \subseteq C) \Rightarrow A \cup B \subseteq C$.

Proof of part (ii). Suppose that $A \subseteq C$ and $B \subseteq C$ $(C \subseteq S)$.

Let x be an arbitrary element of $A \cup B$. Then x belongs to at least one of A, B. If $x \in A$, then also $x \in C$ (since $A \subseteq C$). Otherwise, $x \in B$ and so again $x \in C$ (since $B \subseteq C$). So in all cases $x \in C$.

This shows that: $x \in A \cup B \Rightarrow x \in C$, i.e. that $A \cup B \subseteq C$.

Part (ii) is now proved.

It is possible (and slightly easier) to prove part (i) by a similar argument.

Although neither part of 7.2 can be regarded as a profound result, the format of the simple proof given of part (ii) is very worthy of study by newcomers to the art of rigorous argument about sets.

A further pair of easy results that can be deduced from 7.2 is:

7.3 For sets A, B,
  (i) $A \cup B = B \Leftrightarrow A \subseteq B$,   (ii) $A \cap B = B \Leftrightarrow B \subseteq A$.

Finally, we mention a simple version of the *distributive laws* for sets, viz:

7.4 For sets A, B, C,
  (i) $A \cap (B \cup C) = (A \cap B) \cup (A \cap C)$,
  (ii) $A \cup (B \cap C) = (A \cup B) \cap (A \cup C)$.

Recently it has become fashionable to give prominence in elementary mathematics courses to the technique of proving by Venn diagram results (such as 7.4) about three sets or subsets. An account of the technique may be found in, for example, *Algebra and Number Systems* by J. Hunter *et al.* [2]. The drawback in "proofs by Venn diagram" is that they involve an appeal (of dubious logical status) to a picture. It is left as a challenge to the student to find a proof of 7.4 that involves no such appeal to a picture.

8. Cartesian product of sets

For any positive integer n, an n-**tuple** means something of the form $(x_1, x_2, \ldots, x_n)$, i.e. a sequence of n objects $x_1, x_2, \ldots, x_n$ listed in a definite order. In particular, a 2-tuple is something of the form (x, y) and is usually called an **ordered pair**.

It must be emphasized that the order matters in an n-tuple; e.g. we look upon $(1, 2, 3)$ and $(2, 1, 3)$ as different 3-tuples. In fact, we consider two n-tuples $(x_1, x_2, \ldots, x_n)$, $(y_1, y_2, \ldots, y_n)$ to be equal if and only if $x_i = y_i$ for every i from 1 to n.

Note also that repetitions are allowed in an n-tuple; e.g. $(1, 2, 1)$ and $(1, 1, 1)$ are perfectly good 3-tuples.

Definition. Let $A_1, A_2, \ldots, A_n$ be sets. The **cartesian product** $A_1 \times A_2 \times \ldots \times A_n$ of the given sets (in their given order) is the set of all n-tuples with first object belonging to A_1, second object belonging to $A_2, \ldots$, and nth object belonging to A_n. This may be expressed concisely by writing

$$A_1 \times A_2 \times \ldots \times A_n = \{(x_1, x_2, \ldots, x_n) : x_i \in A_i \quad \text{for} \quad i = 1, 2, \ldots, n\}.$$

In particular, for sets A, B, the cartesian product $A \times B$ is the set of all ordered pairs with first object in A, second object in B.

Remarks. (*a*) For a given set A, we usually denote the cartesian product $A \times A \times \ldots \times A$ (n factors) by A^n. For example, $\mathbb{R}^2$ is the set of all ordered pairs of real numbers.

(*b*) It is natural to identify $\mathbb{R}^2$ with the coordinate plane: each element (a, b) of $\mathbb{R}^2$ is thought of as a point of the coordinate plane (viz. the point with coordinates (a, b)), and $\mathbb{R}^2$, being the set of all such objects, is thought of as the entire coordinate plane.

(*c*) Observe that $(\frac{1}{2}, 1) \in \mathbb{R} \times \mathbb{Z}$, while $(\frac{1}{2}, 1) \notin \mathbb{Z} \times \mathbb{R}$, and therefore $\mathbb{R} \times \mathbb{Z} \neq \mathbb{Z} \times \mathbb{R}$. Thus $A \times B$ and $B \times A$ certainly may be different; and indeed it is not hard to prove that $A \times B \neq B \times A$ whenever A, B are unequal non-empty sets.

(*d*) Clearly, for every set A, $A \times \emptyset = \emptyset \times A = \emptyset$.

(*e*) It is left as an exercise for the reader to prove that:

8.1 If A, B are finite sets, then $|A \times B| = |A| \times |B|$.

Worked example. Let A, B, C be sets with $B \neq \emptyset$. Given that $A \times B \subseteq C \times C$, show that $A \subseteq C$.

Solution. Since $B \neq \emptyset$, we can introduce into the argument an element $b \in B$. We have:

$$x \in A \Rightarrow (x, b) \in A \times B$$
$$\Rightarrow (x, b) \in C \times C \quad \text{(since } A \times B \subseteq C \times C)$$
$$\Rightarrow x \in C.$$

It follows that $A \subseteq C$.

EXERCISES ON CHAPTER ONE

1. Suppose that S is a given subset of $\mathbb{R}$. Write down in simplest form the negations of the statements (i) $\forall x \in S,\ x \leqslant 3$, (ii) $\exists x \in S$ s.t. $x \leqslant 3$.

State whether each of (i), (ii) is true or false (a) when $S = \{2, 3, 4\}$, (b) when $S = [0, 3]$. Which of these four answers can be justified by citing a single value of x in S?

2. Of the statements

(i) $x < 5 \Rightarrow x \leqslant 5$ $(x \in \mathbb{R})$,   (ii) $x \leqslant 5 \Rightarrow x < 5$ $(x \in \mathbb{R})$,

(iii) $x > 5 \Rightarrow x \geqslant 5$ $(x \in \mathbb{R})$,   (iv) $x \geqslant 5 \Rightarrow x > 5$ $(x \in \mathbb{R})$,

(a) which is the converse of (i), (b) which is the contrapositive of (i), and (c) which are true and which false?

3. Suppose that x, y, z are nonzero real numbers, not all equal to each other. Show that x/y, y/z, z/x are not all equal.

4. What is the order (or cardinality) of each of the following sets:

$$S_1 = \{-1, 0, 1\}, \quad S_2 = \{x \in \mathbb{Z} : 0 < x < 6\}, \quad S_3 = \{x^2 - x : x \in S_1\},$$
$$S_4 = \{X \in \mathscr{P}(S_2) : |X| = 3\}, \quad S_5 = \mathscr{P}(\emptyset)?$$

5. Let A, B be subsets of the set S. Prove that $A - B = A \cap \mathscr{C}_S B$, and deduce (without appealing to a Venn diagram) that

$$A - (A - B) = B - (B - A) = A \cap B.$$

6. Let A, B denote subsets of the set S. Write down a simple form of the negation of "$\forall x \in A,\ x \in \mathscr{C}_S B$". Hence prove that, for $A, B \subseteq S$,

$$A \cap B = \emptyset \Leftrightarrow A \subseteq \mathscr{C}_S B.$$

Prove further that

$$\exists X \in \mathscr{P}(S) \text{ s.t. } (A \cap X) \cup (B \cap X') = \emptyset \Leftrightarrow A \cap B = \emptyset$$

(where X' denotes $\mathscr{C}_S X$).

7. For sets A, B, prove that (i) $\mathscr{P}(A) \cup \mathscr{P}(B) \subseteq \mathscr{P}(A \cup B)$ and (ii) $\mathscr{P}(A) \cap \mathscr{P}(B) = \mathscr{P}(A \cap B)$. Give an example of a pair of sets C, D such that $\mathscr{P}(C) \cup \mathscr{P}(D) \neq \mathscr{P}(C \cup D)$.

8. Let A, B be finite sets. What can you say about the relation between $|A \cup B|$ and $|A| + |B|$ (a) when A, B are disjoint, (b) when A, B are not disjoint?

9. The *symmetric difference* $A + B$ of two subsets A, B of the set S is defined by $A + B = (A - B) \cup (B - A)$.

(i) Simplify $A + \emptyset$, $A + S$, $A + A$, $A + \mathscr{C}_S A$ $(A \subseteq S)$.

(ii) Prove that, for $A, B, C \subseteq S$, $(A + B) + C = A + (B + C)$.

10. For each $n \in \mathbb{N}$, let $A_n = [0, 1/n]$. Show that the intersection of the infinite family of sets $A_1, A_2, A_3, \ldots$ is $\{0\}$.

11. Let A, B, C, D be subsets of a set S, and let $P = A \cap B \cap (C \cup D)$, $Q = (A \cap D) \cup (B \cap C)$. By using 7.4 to re-express P, prove that $P \subseteq Q$. Prove also that $P = Q$ iff both $B \cap C \subseteq A$ and $A \cap D \subseteq B$.

12. Let A, B, C, D be subsets of a set S. Show that if there is a subset X of S satisfying the simultaneous equations

$$A \cup (X \cap B) = C, \quad (A \cup X) \cap B = D, \tag{1}$$

then

$$A \cap B \subseteq D \subseteq B \quad \text{and} \quad A \cup D = C. \tag{2}$$

Prove conversely that if the conditions (2) are satisfied, then the equations (1) hold if $X = D - A$.

13. Let A, B be non-empty sets. Prove that if $A \times B = B \times A$, then $A = B$.

14. For sets A, B, C, D, prove that $(A \times B) \cap (C \times D) = (A \cap C) \times (B \cap D)$.

15. Show that, for sets P, Q, R, $P \times (Q \cup R) = (P \times Q) \cup (P \times R)$.
Deduce that, for sets A, B, C, D, $(A \times B) \cup (C \times D) \subseteq (A \cup C) \times (B \cup D)$, and give an example to show that equality need not occur.

16. How many partitions are there of the set S (i) if $|S| = 2$, (ii) if $|S| = 3$, (iii) if $|S| = 4$?

CHAPTER TWO

SOME PROPERTIES OF $\mathbb{Z}$

9. Introduction

One of our aims in this chapter is to establish certain elementary facts about the system of integers that we need in order to develop and illustrate the material of later chapters. These facts include simple results on congruences (introduced in §14) and, on a more fundamental level, the well-ordering principle (an absolutely basic property of $\mathbb{Z}$, explained in detail in §10) and the division algorithm (§11). The last of these is the fact, suggested by our experience of elementary arithmetic, that, given an integer a and a divisor d ($\in \mathbb{N}$), we can, in exactly one way, express a in the form

$$a = (\text{QUOTIENT}) \times d + (\text{REMAINDER})$$

where the quotient and remainder are integers and the remainder lies in the range 0 to $d-1$, inclusive.

This chapter also includes proof of the very straightforward factorization properties of the integers (see §13). The student is advised to note particularly how the proof given of these properties depends crucially on the important result 12.3—the fact that the highest common factor of two positive integers a, b can be expressed in the form $x_1 a + x_2 b$ (with $x_1, x_2 \in \mathbb{Z}$). In abstract algebra, this material about $\mathbb{Z}$ motivates fruitful exploration of factorization in a more general framework, where one discovers, in particular, that many other algebraic systems have nice factorization properties like those of $\mathbb{Z}$. We return to this topic in the final chapter of the book.

10. The well-ordering principle

We first introduce some terminology. In the definitions which follow, S will denote a subset of $\mathbb{R}$.

22

Definition 1. We say that S is **bounded below** iff $\exists\, k \in \mathbb{R}$ such that

$$\forall x \in S, \quad k \leqslant x. \qquad (\alpha)$$

And when S is bounded below, each number k that satisfies (α) is called a **lower bound** of S.

[For example, the interval $U = (1, \infty)$ is bounded below; -7, 1, $\frac{1}{2}$ are examples of lower bounds of U; the set of all lower bounds of U is $(-\infty, 1]$.]

Definition 2. We say that S is **bounded above** iff $\exists\, K \in \mathbb{R}$ such that

$$\forall x \in S, \quad K \geqslant x. \qquad (\beta)$$

And when S is bounded above, each number K satisfying (β) is called an **upper bound** of S.

[For example, the interval $V = (-\infty, 2]$ is bounded above; the set of all the upper bounds of V is $[2, \infty)$.]

We are now in a position to state one form of the well-ordering principle, namely:

10.1 Every non-empty set of integers that is bounded below has a least member, i.e. a member which is less than or equal to every member of the set.

A dual version is:

10.2 Every non-empty set of integers that is bounded above has a greatest member, i.e. a member which is greater than or equal to every member of the set.

It should be noted that a non-empty subset of $\mathbb{R}$ that is bounded below need not have a least member: e.g. the interval $(1, \infty)$, though bounded below, clearly does not have a least member. This observation highlights the fact that 10.1 says something distinctive and non-trivial about the integers (as does 10.2).

It is also true that the well-ordering principle is a fundamental property of $\mathbb{Z}$, in the sense that no description of $\mathbb{Z}$ is adequate unless it incorporates the well-ordering principle or something equivalent to it (e.g. the principle of induction). Accordingly, any so-called proof of the well-ordering principle is bound to be misleading to the extent that it must somewhere appeal to an assumption that is no more fundamental than the well-ordering principle. [Students familiar with the principle of induction can gain some sense of perspective in these matters by thinking out how the well-ordering principle can be proved from the principle of induction and vice versa.]

In a number of proofs we shall make use of the following consequence of 10.1 which may be termed the principle of the least counterexample.

10.3 Suppose that $P(n)$ is a condition involving the variable n, whose domain is the set $\{n \in \mathbb{Z} : n \geqslant l\}$, l being a fixed integer. Suppose further that the universal statement "$\forall n \geqslant l$, $P(n)$" is false. Then there is a least counter-example, i.e. a lowest integral value ($\geqslant l$) of n for which $P(n)$ is false.

Proof. Let S be the set of all counterexamples to the universal statement, i.e. the set of all values ($\geqslant l$) of n for which $P(n)$ is false. Then, since the universal statement is false, $S \neq \emptyset$. Moreover, S consists of integers and is bounded below (all its members being $\geqslant l$). By 10.1, it follows that S has a least member; and this proves the result.

Integral part of a real number

Let $x \in \mathbb{R}$. We shall take it for granted that there exist integers which are $\leqslant x$. So the set of all such integers is non-empty and is bounded above (x being an upper bound). Accordingly (by 10.2) there is a greatest integer less than or equal to x: this integer is called the **integral part** of x and is denoted by $[x]$, e.g.

$$[\tfrac{5}{2}] = 2, \quad [-\tfrac{5}{2}] = -3, \quad [14] = 14, \quad [\sqrt{10}] = 3.$$

Certain things follow immediately from the definition:

10.4 Let $x \in \mathbb{R}$. Then:
   (i) $[x] \leqslant x < [x] + 1$;
   (ii) $x - [x] \in [0, 1)$;
   (iii) if $x = n + t$, where $n \in \mathbb{Z}$ and $t \in [0, 1)$, then $n = [x]$.

11. The division algorithm

The formal statement of this result, briefly previewed in §9, is:

11.1 Let a, d be integers, with $d > 0$. Then there exist unique integers q, r such that

$$a = qd + r \quad \text{and} \quad 0 \leqslant r \leqslant d - 1. \tag{*}$$

Proof. Part (i)—the existence of q, r. Here we shall produce integers q, r satisfying (*).

Let $q = \left[\dfrac{a}{d}\right]$ and $t = \dfrac{a}{d} - \left[\dfrac{a}{d}\right]$, so that $q \in \mathbb{Z}$ and $t \in [0, 1)$. Since $a/d = q + t$, we have

$$a = qd + r \tag{**}$$

where $r = td$. Since $a, q, d \in \mathbb{Z}$, it follows from (**) that r ($= a - qd$) is an integer too. Further, since $0 \leqslant t < 1$ and $d > 0$, it follows that $0 \leqslant td < d$, i.e. $0 \leqslant r < d$; and hence, since $r \in \mathbb{Z}$, $0 \leqslant r \leqslant d - 1$.

We have thus produced q, r with the stated properties.

Part (ii)—the uniqueness of q, r.

Suppose that $a = qd + r$, where $q, r \in \mathbb{Z}$ and $0 \leqslant r \leqslant d - 1$. We have to prove that these suppositions uniquely determine q, r in terms of a and d. We have

$$\frac{a}{d} = q + \frac{r}{d} \quad \text{and} \quad 0 \leqslant \frac{r}{d} \leqslant \frac{d-1}{d} < 1.$$

Hence (cf. 10.4(iii)) $q = \left[\dfrac{a}{d}\right]$, and so $r = a - d\left[\dfrac{a}{d}\right]$.

The result follows.

In the above result, we call r the **principal remainder** on division of a by d.

If and only if $r = 0$, i.e. iff a is an integral multiple of d, we say that d **divides** a (or, alternatively, d is a **factor** of a). We often use a short vertical line to stand for the verb "divides", so that

$$d|a, \quad d \nmid a$$

are the standard shorthands for, respectively, "d divides a" and "d does not divide a". So, for example, $5|(-30)$ and $4 \nmid 19$.

Simple results on this theme include:

11.2 If $d|a$, where $d \in \mathbb{N}$ and a is a nonzero integer, then $|a| \geqslant d$.

11.3 If $d|e$ and $e|f$, where $d, e, f \in \mathbb{N}$, then $d|f$.

11.4 If $d|a$ and $d|b$, where $d, a, b \in \mathbb{Z}$ and $d > 0$, then $d|(a+b)$ and $d|(a-b)$ and, more generally, $d|(x_1 a + x_2 b)$ for every $x_1, x_2 \in \mathbb{Z}$.

The proofs of these results are straightforward. As an indication of method of attack, here is a proof of 11.4.

Proof. Suppose that $d|a$ and $d|b$ ($d, a, b \in \mathbb{Z}, d > 0$). Let x_1, x_2 be arbitrary integers.

Since $d|a$ and $d|b$, $a = kd$ and $b = ld$ for some integers k, l. Hence

$$x_1 a + x_2 b = x_1 kd + x_2 ld = (x_1 k + x_2 l)d.$$

Since $x_1 k + x_2 l \in \mathbb{Z}$, it follows that $d|(x_1 a + x_2 b)$.

This proves 11.4.

12. Highest common factors and Euclid's algorithm

Let $a_1, a_2, \ldots, a_n$ be a finite sequence of positive integers. Each positive integer which divides all of $a_1, a_2, \ldots, a_n$ is called a **common factor** of $a_1, a_2, \ldots, a_n$ (e.g. 3 is a common factor of 12, 18, 30). In the general case there is certainly at least one common factor of $a_1, a_2, \ldots, a_n$, viz. 1. On the other hand, every such common factor is, in particular, a factor of a_1 and so (cf. 11.2) is not greater than a_1. Clearly, therefore, the set of common factors of $a_1, a_2, \ldots, a_n$ has a greatest member, which is termed the **highest common factor** of $a_1, a_2, \ldots, a_n$

and denoted by

$$(a_1, a_2, \ldots, a_n) \quad \text{or} \quad \text{hcf}(a_1, a_2, \ldots, a_n).$$

We call two positive integers a, b **relatively prime** iff their highest common factor is 1, e.g. the integers 6 and 25 are relatively prime (even though neither is a prime number). [The student should know that a **prime number** means an integer $p \geqslant 2$ whose only factors (in $\mathbb{N}$) are 1 and p. An integer greater than 1 that is not prime is termed **composite**.]

It is easily seen that:

12.1 If p is a prime number and a is a positive integer such that $p \nmid a$, then p and a are relatively prime.

We now focus attention on the determination of $\text{hcf}(a_1, a_2)$, where a_1, a_2 are given positive integers and where, let us suppose, $a_1 \geqslant a_2$. The highest common factor may be found systematically by a process called *Euclid's algorithm*, exhibited below. This process, it will be seen, gives a very useful by-product—the fact that the highest common factor of a_1 and a_2 can be expressed as an integral linear combination of a_1 and a_2, i.e. in the form $x_1 a_1 + x_2 a_2$ with x_1, x_2 integers.

Starting from the given positive integers a_1, a_2, we can make repeated use of the division algorithm to obtain (all symbols denoting integers):

$$a_1 = q_1 a_2 + a_3, \quad \text{where} \quad 0 \leqslant a_3 < a_2 (\leqslant a_1);$$
$$\text{and, if } a_3 \neq 0, \quad a_2 = q_2 a_3 + a_4, \quad \text{where} \quad 0 \leqslant a_4 < a_3 < a_2 \leqslant a_1;$$
$$\text{and, if } a_4 \neq 0, \quad a_3 = q_3 a_4 + a_5, \quad \text{where} \quad 0 \leqslant a_5 < a_4 < a_3 < a_2 \leqslant a_1;$$
$$\text{etc.}$$

Since there is only a finite number of integers in the interval $[0, a_1]$, we must, after some finite number of steps in the process, obtain zero remainder. Thus we obtain a finite sequence of equations

$$a_1 = q_1 a_2 + a_3 \tag{1}$$
$$a_2 = q_2 a_3 + a_4 \tag{2}$$
$$a_3 = q_3 a_4 + a_5 \tag{3}$$
$$\cdots$$
$$a_{r-3} = q_{r-3} a_{r-2} + a_{r-1} \tag{$r-3$}$$
$$a_{r-2} = q_{r-2} a_{r-1} + a_r \tag{$r-2$}$$
$$a_{r-1} = q_{r-1} a_r \tag{$r-1$}$$

12.2 In this process a_r (which is the last nonzero remainder, except in the trivial case $a_2 | a_1$) equals $\text{hcf}(a_1, a_2)$.

Proof. The result is trivial if $a_2|a_1$ (i.e. $r = 2$). Henceforth we deal with the non-trivial case $a_2 \nmid a_1$ (so that $r \geqslant 3$).

Let $d = \text{hcf}(a_1, a_2)$. Using the general version of 11.4, we reason progressively as follows.

By (1), since $d|a_1$ and $d|a_2$, d also divides a_3 ($= a_1 - q_1 a_2$); hence, by (2), since $d|a_2$ and $d|a_3$, d also divides a_4 ($= a_2 - q_2 a_3$); etc; and eventually, at the $(r-2)$th such step : hence, by $(r-2)$, since $d|a_{r-2}$ and $d|a_{r-1}$, d also divides a_r.

Consequently, by 11.2, $a_r \geqslant d$.

But, using the equations in reverse order, we can argue thus:

by $(r-1)$, $a_r|a_{r-1}$;

hence, by $(r-2)$, a_r also divides a_{r-2};

hence, by $(r-3)$, since $a_r|a_{r-1}$ and $a_r|a_{r-2}$, a_r also divides a_{r-3};

etc; and eventually:

hence, by (2), since $a_r|a_4$ and $a_r|a_3$, a_r also divides a_2;

hence, by (1), since $a_r|a_3$ and $a_r|a_2$, a_r also divides a_1.

By the last two conclusions, a_r is a common factor of a_1 and a_2, and so $a_r \leqslant d$ (d being the highest such common factor).

Since it has already been proved that $a_r \geqslant d$, it follows that $a_r = d$, which is the stated result.

At this stage it is instructive to work through a specific numerical example. Let us consider $\text{hcf}(2190, 465)$.

Following the procedure described, we successively obtain

$$2190 = 4.465 + 330 \tag{1}$$
$$465 = 1.330 + 135 \tag{2}$$
$$330 = 2.135 + 60 \tag{3}$$
$$135 = 2.60 + 15 \tag{4}$$
$$60 = 4.15 \tag{5}$$

By 12.2, we conclude that $\text{hcf}(2190, 465) = 15$.

More information lies close at hand. By using the equations (4), (3), (2), (1) in that order, we can express the highest common factor 15 as an integral linear combination of 2190 and 465. We proceed as follows.

$15 = 135 - 2.60$ (by (4))

$\quad = 135 - 2(330 - 2.135)$ (using (3) to replace the smaller of 60, 135 in the previous line)

$\quad = 5.135 - 2.330$

$\quad = 5(465 - 330) - 2.330$ (using (2) to replace the smaller of 135, 330 in the previous line)

$\quad = 5.465 - 7.330$

$\quad = 5.465 - 7(2190 - 4.465)$ (using (1) to replace the smaller of 330, 465 in the previous line)

$\quad = 33.465 - 7.2190.$

Now consider again an arbitrary pair of positive integers a_1, a_2 with $a_1 \geqslant a_2$. Provided $a_2 \nmid a_1$, it is evident that the process demonstrated above can be carried out to yield an expression for $\text{hcf}(a_1, a_2)$ as an integral linear combination of a_1, a_2; and in the trivial case $a_2 | a_1$ we of course have

$$\text{hcf}(a_1, a_2) = a_2 = 0a_1 + 1a_2$$

Therefore:

12.3 If $a, b \in \mathbb{N}$ and $d = \text{hcf}(a, b)$, then $\exists x_1, x_2 \in \mathbb{Z}$ such that

$$d = x_1 a + x_2 b.$$

This theorem has important consequences, including the results 12.4, 12.5 below.

12.4 Let $a, b \in \mathbb{N}$, let $d = \text{hcf}(a, b)$, and let c be any common factor of a and b. Then $c|d$.

Proof. By 12.3, $d = x_1 a + x_2 b$ for some $x_1, x_2 \in \mathbb{Z}$. Since $c|a$ and $c|b$, it follows by 11.4 that $c|d$.

It is of interest that 12.4 shows that the highest common factor of two integers is supreme among the common factors of the two integers in two senses. As well as being supreme in the sense of numerically greatest, it is also supreme in the sense of being divisible by every common factor. This observation gives a clue as to how one can introduce the concept of "highest common factor" in other algebraic systems where "numerically greatest" would be meaningless.

12.5 Suppose that the prime number p divides the product ab, where $a, b \in \mathbb{N}$. Then $p|a$ or $p|b$ (possibly both).

Proof. Suppose that $p \nmid a$. (It will suffice to prove that $p|b$ follows from this assumption.)

By 12.1, $\text{hcf}(p, a) = 1$ and hence, by 12.3, there are integers x_1, x_2 such that $x_1 p + x_2 a = 1$. Hence we have

$$b = b \times 1 = b(x_1 p + x_2 a) = (bx_1)p + x_2(ab),$$

which (by 11.4) is divisible by p (since $p|ab$).

Thus $p|b$, and the result follows.

A generalization of 12.5 is obtainable by repeated application:

12.6 If the prime number p divides the product $a_1 a_2 \ldots a_k$ (where each $a_i \in \mathbb{N}$), then p divides at least one of $a_1, a_2, \ldots, a_k$.

Proof. $p|a_1 a_2 a_3 \ldots a_k \Rightarrow p|a_1$ or $p|a_2 a_3 \ldots a_k$ (by 12.5)
$$\Rightarrow p|a_1 \text{ or } p|a_2 \text{ or } p|a_3 \ldots a_k \text{ (by 12.5)}$$
$$\ldots$$
$$\Rightarrow p|a_1 \text{ or } p|a_2 \text{ or } p|a_3 \text{ or } \ldots \text{ or } p|a_k,$$
after altogether $k-1$ applications of 12.5.

The basic results 12.3 and 12.4 are also capable of generalization. We state the more general results without proof.

12.7 Let $a_1, a_2, \ldots, a_n$ be positive integers, and let d be their highest common factor. Then:
   (i) d can be expressed as $x_1 a_1 + x_2 a_2 + \ldots + x_n a_n$ for some $x_1, x_2, \ldots, x_n \in \mathbb{Z}$;
   (ii) every common factor of $a_1, a_2, \ldots, a_n$ is a divisor of d.

13. The fundamental theorem of arithmetic

The *fundamental theorem of arithmetic* is the name given to the twofold assertion that, for every integer $n \geqslant 2$,
   (i) n can be expressed as a product of one or more primes, and
   (ii) there is essentially only one expression for n as a product of primes, i.e. any two such expressions for n can differ only in the order in which the prime factors appear.
   (Note that we allow the simple case in which there is just one factor in the "product". Indeed if p is a prime number, the one and only way to express p as a product of primes is to write it as the one-factor product p.)
   The unique expression for the integer n ($\geqslant 2$) as a product of primes is called the **prime decomposition** of n. Examples of prime decompositions include:
$$504 = 2 \times 2 \times 2 \times 3 \times 3 \times 7, \quad 16 = 2 \times 2 \times 2 \times 2, \quad 17 = 17,$$
$$1978 = 2 \times 23 \times 43, \quad 1988 = 2 \times 2 \times 7 \times 71.$$

The results 13.1 and 13.2, which follow, establish respectively parts (i) and (ii) of the fundamental theorem (as stated above).

13.1 For every integer $n \geqslant 2$, n can be expressed as a product of primes.

Proof. With a view to obtaining a contradiction, suppose the result false. Then, by 10.3, there is a least counterexample integer m ($\geqslant 2$) which cannot be expressed as a product of primes. As a prime number is, in a trivial way, expressible as a product of primes, it is clear that m cannot be prime. Hence $m = kl$ for some $k, l \in \mathbb{N}$ satisfying $1 < k, l < m$. Because m is the least counterexample, both k and l are expressible as products of primes: say $k = p_1 p_2 \ldots p_r$, $l = q_1 q_2 \ldots q_s$ (all of $p_1, \ldots, q_s$ being primes). Hence we have
$$m = p_1 p_2 \ldots p_r q_1 q_2 \ldots q_s,$$

giving an expression for the supposed counterexample m as a product of primes. This is a contradiction, and from it the result follows.

13.2 For every integer $n \geqslant 2$, any two expressions for n as products of primes differ only in the order of the factors.

Proof. Suppose, with a view to obtaining a contradiction, that the result is false. Then, by 10.3, there is a least counterexample integer m which can be expressed as a product of primes in two genuinely different ways, say

$$m = p_1 p_2 \ldots p_r = q_1 q_2 \ldots q_s,$$

where $p_1, \ldots, p_r, q_1, \ldots, q_s$ are all primes. Clearly m itself is not prime and so $r, s \geqslant 2$. Without loss of generality we can insist that, in each of the prime products equal to m, the prime factors are arranged in ascending order, i.e. $p_1 \leqslant p_2 \leqslant \ldots \leqslant p_r$ and $q_1 \leqslant q_2 \leqslant \ldots \leqslant q_s$; and also that $p_1 \leqslant q_1$.

We note that $p_1 \neq q_1$; for, if p_1 were equal to q_1, we would have

$$\frac{m}{p_1} = p_2 p_3 \ldots p_r = q_2 q_3 \ldots q_s$$

giving two genuinely different prime product expressions for the integer m/p_1, which, however, is less than the least counterexample m.

Hence $p_1 < q_1$, and so, since $q_1 \leqslant q_2 \leqslant \ldots \leqslant q_s$, p_1 is less than q_j for each j. Now each q_j is prime and so, since $1 < p_1 < q_j$, we deduce that, for each j, $p_1 \nmid q_j$.

But p_1 does divide m, which equals $q_1 q_2 \ldots q_s$, and so, by 12.6, p_1 must divide at least one of $q_1, q_2, \ldots, q_s$.

We thus have a contradiction, and the stated result follows.

14. Congruence modulo m $(m \in \mathbb{N})$

When we face a question like, "If it is now 8 o'clock, what time will it be 30 hours from now?", we find ourselves wanting to ignore integral multiples of 12. This is a somewhat trivial example of a host of reasons why it is profitable to have a special phraseology and notation for indicating that two integers differ by an integral multiple of some given integer m. That is basically what this section is about. Throughout the general discussion, m will denote a fixed positive integer.

For $a, b \in \mathbb{Z}$, we write

$$a \equiv b \pmod{m}$$

(read "a is *congruent* to b modulo m") to mean $m|(b-a)$, i.e. $b-a$ is an integral multiple of m, i.e. $b = a + km$ for some integer k.

So, for example, it is true that $13 \equiv 5 \pmod{4}$ and $9 \equiv -1 \pmod{5}$, while, equally clearly, $10 \not\equiv 2 \pmod 6$, i.e. 10 is not congruent to 2 modulo 6.

It is useful at the outset to note the simple information in the following results 14.1 and 14.2.

14.1 (i) For every integer a, $a \equiv a \pmod m$.

(ii) $a \equiv b \pmod m \Rightarrow b \equiv a \pmod m$ $(a, b \in \mathbb{Z})$.

(iii) $a \equiv b \pmod m$ and $b \equiv c \pmod m \Rightarrow a \equiv c \pmod m$ $(a, b, c \in \mathbb{Z})$.

Proof. (i) is true because, for every $a \in \mathbb{Z}$, $a - a$ is an integral multiple of m (viz. $0 \times m$).

(ii) For $a, b \in \mathbb{Z}$,

$$a \equiv b \pmod m \Rightarrow b = a + km \quad (k \text{ being an integer})$$
$$\Rightarrow a - b = (-k)m$$
$$\Rightarrow m|(a - b) \quad (\text{since } -k \text{ is an integer})$$
$$\Rightarrow b \equiv a \pmod m.$$

(iii) For $a, b, c \in \mathbb{Z}$,

$$a \equiv b \pmod m \text{ and } b \equiv c \pmod m \Rightarrow m|(b - a) \text{ and } m|(c - b)$$
$$\Rightarrow m|\{(b - a) + (c - b)\} \quad (\text{by } 11.4)$$
$$\Rightarrow m|(c - a)$$
$$\Rightarrow a \equiv c \pmod m.$$

14.2 (i) If the integer a has principal remainder r on division by m, then $a \equiv r \pmod m$.

(ii) Each integer is congruent (modulo m) to precisely one of the integers $0, 1, 2, \ldots, m - 1$.

Proof. (i) If a has principal remainder r on division by m, then

$$a = qm + r \text{ for some integer } q,$$

and hence $a \equiv r \pmod m$.

(ii) Let a be an arbitrary integer. It follows at once from (i) that a is congruent (modulo m) to at least one of the integers in the interval $[0, m - 1]$. If a were congruent (modulo m) to two different integers r_1, r_2 in this interval, we would have

$$a = q_1 m + r_1 = q_2 m + r_2 \text{ for some } q_1, q_2 \in \mathbb{Z}$$

—an impossibility in view of the uniqueness clause in the division algorithm (11.1). The result follows.

One of the advantages of handling information like "$m|(b - a)$" in the form "$a \equiv b \pmod m$" is that sentences of the latter form can be treated to a great extent like equations. The next theorem amplifies this claim.

14.3 Suppose that $a \equiv b \pmod{m}$ and that $c \equiv d \pmod{m}$, where $a, b, c, d \in \mathbb{Z}$. Then:

   (i) $a + c \equiv b + d \pmod{m}$;
   (ii) $a - c \equiv b - d \pmod{m}$;
   (iii) $ac \equiv bd \pmod{m}$.

Proof. Since $a \equiv b \pmod{m}$ and $c \equiv d \pmod{m}$,

$$b = a + km \quad \text{and} \quad d = c + lm, \quad \text{for some } k, l \in \mathbb{Z}.$$

(i) $(b + d) - (a + c) = (a + km + c + lm) - (a + c) = (k + l)m$. Since $k + l \in \mathbb{Z}$, it follows that $a + c \equiv b + d \pmod{m}$.

(ii) is proved similarly by considering $(b - d) - (a - c)$.

(iii) $bd = (a + km)(c + lm) = ac + (kc + al + klm)m$. Therefore, since $kc + al + klm \in \mathbb{Z}$, $ac \equiv bd \pmod{m}$.

The conclusions (i), (ii), (iii) in 14.3 may be described as the results of (respectively) adding, subtracting, and multiplying the given congruences (i.e. the given statements of the form $x \equiv y \pmod{m}$).

Notice that certain things can be deduced via 14.3(iii) from one given congruence $a \equiv b \pmod{m}$. By multiplying the given congruence by the clearly true congruence $c \equiv c \pmod{m}$ [c being an arbitrary integer], we deduce that $ca \equiv cb \pmod{m}$. On another line of attack, by multiplying the given congruence by itself, we deduce that $a^2 \equiv b^2 \pmod{m}$; then, multiplying this by the given congruence, we deduce $a^3 \equiv b^3 \pmod{m}$; and, by continuing in this way, we can deduce $a^n \equiv b^n \pmod{m}$ for any specified positive integer n. To sum up:

14.4 If $a, b \in \mathbb{Z}$ and $a \equiv b \pmod{m}$, then
   (i) $\forall c \in \mathbb{Z}, ca \equiv cb \pmod{m}$,
   (ii) $\forall n \in \mathbb{N}, a^n \equiv b^n \pmod{m}$.

Worked examples

1. Show that, for every $n \in \mathbb{Z}$, $n^2 \not\equiv 2$ and $n^2 \not\equiv 3 \pmod{5}$. [Note: this is a fact familiar to the observant—that (in the customary notation for integers) the square of an integer never ends in 2, 3, 7, or 8.]

Solution. Let n be an arbitrary integer. By 14.2(ii),

$$n \equiv 0 \text{ or } n \equiv 1 \text{ or } n \equiv 2 \text{ or } n \equiv 3 \text{ or } n \equiv 4 \pmod{5}.$$

Hence, by 14.4(ii),

$$n^2 \equiv 0 \text{ or } n^2 \equiv 1 \text{ or } n^2 \equiv 4 \text{ or } n^2 \equiv 9 \text{ or } n^2 \equiv 16 \pmod{5}.$$

Since $9 \equiv 4$ and $16 \equiv 1 \pmod{5}$, it follows [by 14.1(iii)] that, in every case, $n^2 \equiv 0$ or 1 or 4 (mod 5).

By 14.2(ii), the result follows.

2. Using the result of example 1, show that there do not exist integers x, y satisfying $7x^2 - 15y^2 = 1$.

Solution. Suppose the contrary, so that $7x^2 - 15y^2 = 1$ for some integers x, y. Then (since $15y^2$ is an integral multiple of 5)

$$7x^2 \equiv 1(\text{mod } 5).$$

Since $7 \equiv 2 \,(\text{mod } 5)$, it follows (by 14.4(i)) that $7x^2 \equiv 2x^2 \,(\text{mod } 5)$ and hence (by 14.1) that $2x^2 \equiv 1 \,(\text{mod } 5)$.

But, by example 1, $x^2 \equiv 0$ or 1 or $4 \,(\text{mod } 5)$ and hence $2x^2 \equiv 0$ or 2 or 8 (mod 5), i.e. $2x^2 \equiv 0$ or 2 or 3 (mod 5).

In view of 14.2(ii), we now have contradictory information about the integer $2x^2$. From this contradiction the result follows.

EXERCISES ON CHAPTER TWO

1. Let $x \in \mathbb{R}$. Prove that $[x] + [-x]$ equals 0 if $x \in \mathbb{Z}$ and equals -1 if $x \notin \mathbb{Z}$.

2. Suppose that a, b are relatively prime positive integers and that the positive integer c is a factor of $a + b$. Prove that $\mathrm{hcf}(a, c) = \mathrm{hcf}(b, c) = 1$.

3. Show that, $\forall n \in \mathbb{N}$, $\mathrm{hcf}(9n + 8, 6n + 5) = 1$.

4. Let a, b, c denote positive integers, and let d denote $\mathrm{hcf}(a, b)$. Prove that
$$\exists\, x,\, y \in \mathbb{Z} \ \text{s.t.}\ ax + by = c \Leftrightarrow d \,|\, c.$$

5. Suppose that d, a, b are positive integers, that a and d are relatively prime, and that $d|ab$. Prove that $d|b$.

6. Suppose that d_1, d_2 are relatively prime positive integers such that $d_1|a$ and $d_2|a$, a being a positive integer. Use 12.3 to prove that $d_1 d_2 | a$.

7. Show that $n \equiv 7 \ (\mathrm{mod}\ 12) \Rightarrow n \equiv 3 \ (\mathrm{mod}\ 4) \ (n \in \mathbb{Z})$.

8. Show that every positive integer (when written in the customary notation) is congruent modulo 9 to the sum of its digits. Deduce a simple test for divisibility by 9.

9. Show that, $\forall n \in \mathbb{Z}$, $n^2 \equiv 0$ or 1 (mod 3). Show further that
$$3 \,|\, (x^2 + y^2) \Rightarrow 3|x \text{ and } 3|y \qquad (x, y \in \mathbb{Z}).$$
Deduce that there is no solution in integers of the equation
$$x^2 + y^2 = 3z^2$$
except $x = y = z = 0$.

10. Show that, $\forall n \in \mathbb{N}$, $3^{3n+1} \equiv 3.5^n \ (\mathrm{mod}\ 11)$. Obtain a corresponding result about 2^{4n+3} and deduce that, $\forall n \in \mathbb{N}$, $11|(3^{3n+1} + 2^{4n+3})$.

11. Use Euclid's algorithm to find an integer x such that $31x \equiv 1 \ (\mathrm{mod}\ 56)$.

12. Show that: $a \equiv b \ (\mathrm{mod}\ m) \Rightarrow \mathrm{hcf}(a, m) = \mathrm{hcf}(b, m) \quad (a, b, m \in \mathbb{N})$.

13. An integer m is called a *perfect square* iff $\exists n \in \mathbb{Z}$ such that $m = n^2$, i.e. iff $m \in \{1, 4, 9, 16, 25, \ldots\}$. Suppose that the integer r is a perfect square and that $r = st$, where s, t are relatively prime positive integers. Prove that s and t are perfect squares.

14. Prove that: $\sqrt{n} \in \mathbb{Q} \Rightarrow \sqrt{n} \in \mathbb{N} \quad (n \in \mathbb{N})$.

15. Prove that the set of prime numbers is infinite. [Suggested method: Suppose the contrary, and let $p_1, p_2, \ldots, p_n$ be a complete list of all the primes. Obtain a contradiction by showing that no prime divides the integer $p_1 p_2 \ldots p_n + 1$.]

Prove by a similar approach that there are infinitely many primes that are congruent to 3 (mod 4).

CHAPTER THREE

EQUIVALENCE RELATIONS AND-
EQUIVALENCE CLASSES

15. Relations in general

Underlying the discussion which we now begin is the widespread occurrence in mathematics of sentences of the form

x is related [in some particular way] to y

in which x, y denote elements of some given set S. Examples in the case $S = \mathbb{Z}$ include:

$$x = y; \quad x < y; \quad x \equiv y \,(\text{mod } 5).$$

All such sentences can be written in mathematical shorthand as

$$xRy$$

where R stands for "is related [in some particular way] to": e.g. R could stand for "equals" or "is less than" or "is congruent modulo 5 to".

Naturally, if in any specific context we are going to employ such a shorthand, we must make clear what exactly R stands for. When this has been made fully clear, i.e. when, for any given a, b belonging to the set S, it is possible to decide whether "aRb" is true or false, we say that R is a **relation** on S. So, for instance,

$$= [\text{equality}]$$
$$<$$
$$\equiv (\text{mod } 5) \quad [\text{congruence modulo 5}]$$

are all examples of relations on $\mathbb{Z}$.

When R is a relation on S and x, $y \in S$, a useful notation for the negation of "xRy" is

$$xR'y.$$

Of course, symbols other than R (e.g. ρ, $\sim$) may be used for a relation in general discussions.

It is possible to give a more formal definition of "a relation on S", S being a set. Newcomers to the topic are advised, for the time being at least, to ignore this possibility. However, some students will wish to know that in the formal approach one begins by defining "a relation R on S" to be a subset of $S \times S$; then one introduces "xRy" as an alternative notation for $(x, y) \in R$. It is left to interested students to think out why this approach makes sense and what advantages it has over the informal approach.

The next two sections of the chapter are devoted to the study of equivalence relations, i.e. relations with certain special properties. Knowledge gained in these sections is applied later in the chapter to develop a very significant seam of theory that arises from the relation of congruence (modulo a given positive integer m) on $\mathbb{Z}$.

16. Equivalence relations

Throughout this section, S will denote a set and R will denote a relation on S.

We begin by detailing some properties that R may possess.

(a) We say that R is **reflexive** (or R possesses the reflexive property) iff $\forall x \in S$, xRx.

(b) We say that R is **symmetric** (or R possesses the symmetric property) iff

$$xRy \Rightarrow yRx \qquad (x, y \in S).$$

(c) We say that R is **transitive** (or R possesses the transitive property) iff

$$xRy \text{ and } yRz \Rightarrow xRz \qquad (x, y, z \in S).$$

[Note: when R is symmetric, it is true that

$$xRy \Leftrightarrow yRx \qquad (x, y \in S);$$

for the symmetric property tells us that, if a, $b \in S$ and one of the statements "aRb", "bRa" is true, then so must be the other.]

The main definition of this chapter is:

The relation R on the set S is called an **equivalence relation** iff it is reflexive, symmetric, and transitive.

Obviously "$=$" (equality) is an equivalence relation on S (whatever the set S is). Many less trivial examples are also encountered in elementary mathematics: e.g. in geometry, congruence is an equivalence relation on the set of triangles in a plane. Further, as the result 14.1 tells us:

16.1 For each $m \in \mathbb{N}$, congruence modulo m is an equivalence relation on $\mathbb{Z}$.

Worked example. Let S be the set of all $n \times n$ matrices with real entries. A relation $\sim$ (called *similarity*) is defined on S by

$$X \sim Y \Leftrightarrow \exists \text{ nonsingular } P \in S \text{ s.t. } Y = P^{-1}XP \qquad (X, Y \in S).$$

Prove that $\sim$ is an equivalence relation on S.

Remark. Students must beware of mis-interpreting "$\exists$" here. It is important to understand that, if A, $B \in S$, "$A \sim B$" means that a true statement can be made by inserting some nonsingular matrix in S (depending on what exactly A, B are) into both blanks in the sentence "$B = (..)^{-1}A(..)$"; but the definition of $\sim$ does *not* say that there is a special matrix P that can be used to fill the blanks in all cases.

Solution. (1) For each $X \in S$, $X \sim X$ because $X = I^{-1}XI$ (and I is a nonsingular member of S). Thus $\sim$ is reflexive.

(2) For X, $Y \in S$,

$$X \sim Y \Rightarrow Y = P^{-1}XP \qquad (P \text{ being a nonsingular member of } S)$$
$$\Rightarrow X = PYP^{-1}$$
$$\Rightarrow X = Q^{-1}YQ \qquad (\text{where } Q = P^{-1})$$
$$\Rightarrow Y \sim X \qquad (\text{since } Q \text{ is a nonsingular member of } S).$$

Thus $\sim$ is symmetric.

(3) For X, Y, $Z \in S$,

$$X \sim Y \text{ and } Y \sim Z \Rightarrow Y = P^{-1}XP \quad \text{and} \quad Z = Q^{-1}YQ$$
$$(P, Q \text{ being nonsingular members of } S)$$
$$\Rightarrow Z = Q^{-1}P^{-1}XPQ$$
$$\Rightarrow Z = T^{-1}XT \qquad (\text{where } T = PQ)$$
$$\Rightarrow X \sim Z \qquad (\text{since } T \text{ is a nonsingular member of } S).$$

Thus $\sim$ is transitive.

It now follows that $\sim$ is an equivalence relation on S.

17. Equivalence classes

Throughout this section, R will denote an equivalence relation on the set S.

Definition. Let x be an arbitrary element of S. Each element t of S that satisfies tRx (or equivalently xRt) is called an R-**relative** of x. And the set of all R-relatives of x is called the R-**class** of x (or the **equivalence class** of x *under* R). We shall denote it by C_x. Thus

$$C_x = \{t \in S : tRx\}.$$

For example, if $S = \mathbb{Z}$ and R is congruence modulo 5, then

$$C_0 = \{0, 5, -5, 10, -10, 15, -15, 20, \ldots\}.$$

It is essential to have a thorough understanding of the basic properties of equivalence classes before attempting to study in depth abstract algebra (or indeed any other part of advanced pure mathematics). The rest of this section is mainly devoted to establishing these basic properties, including (17.5 below) the very important fact that the collection of subsets of S that occur as R-classes is a partition of S (or, as some may prefer to express it, the different R-classes form a partition of S). In trying to digest these results, students may find it helpful to think through what each tells us in a very simple case. Consider for example the case where S is a set of coloured beads and the relation R is defined on S by: xRy iff x and y have the same colour $(x, y \in S)$. Clearly this relation R is an equivalence relation, and, for each $x \in S$, the R-class C_x is the set of all beads in S with the same colour as x. So, in this case, theorem 17.5, mentioned above, tells us the fairly obvious fact that the subsets of S each comprising all the beads of a particular colour form a partition of S.

Starting now our exploration of the general case (where R is an arbitrary equivalence relation on the set S), we note first that, because R is reflexive, each element x of S is an R-relative of itself. Therefore:

17.1 For every $x \in S$, C_x is a non-empty subset of S, x being a member of C_x.

The next few results concern the relationship between the subsets C_x and C_y $(x, y \in S)$ (i) when xRy, (ii) when $xR'y$.

17.2 $xRy \Rightarrow C_x = C_y$ $(x, y \in S)$.

Proof. Suppose that xRy.
Then we have, for $t \in S$,

$$\begin{aligned} t \in C_x &\Rightarrow tRx && \text{(by definition of } C_x) \\ &\Rightarrow tRy && \text{(since } xRy \text{ and } R \text{ is transitive)} \\ &\Rightarrow t \in C_y && \text{(by meaning of } C_y). \end{aligned}$$

Hence $C_x \subseteq C_y$.

Just as we have deduced $C_x \subseteq C_y$ from xRy, so from yRx (true also by the symmetric property of R) we can deduce $C_y \subseteq C_x$. Therefore $C_x = C_y$.
The result follows.

It is not difficult to improve 17.2 to:

17.3 $C_x = C_y \Leftrightarrow xRy$ $(x, y \in S)$.

Proof. We know from 17.2 that $xRy \Rightarrow C_x = C_y$ $(x, y \in S)$.
To prove the converse, suppose that $C_x = C_y$ $(x, y \in S)$. Then since (by 17.1) $x \in C_x$, it follows that $x \in C_y$, i.e. xRy. The result now follows.

Remark. It is now clear that C_a and C_b may be the same set when $a \neq b$ $(a, b \in S)$; and it is often important to bear in mind that an equivalence class is liable to have several possible labels of the form C_x.

17.4 If $xR'y$, then $C_x \cap C_y = 0$ $(x, y \in S)$.

Proof. Suppose that $xR'y$ $(x, y \in S)$.

If $C_x \cap C_y$ were non-empty, there would be an element z belonging to both C_x and C_y, so that we would have both xRz and zRy and hence (by the transitive property of R) xRy—a contradiction.

The result follows.

We reach now the climax of this sequence of theorems.

17.5 The set of R-classes is a partition of S.

Proof. By the definition of "partition" (see §7), we have to prove:
  (i) the union of the R-classes is the whole of S;
  (ii) unequal R-classes are disjoint;
  (iii) each R-class is non-empty.

Of these, (iii) and (i) follow from 17.1, which tells us both that every R-class is non-empty and that every element of S belongs to at least one R-class (since $x \in C_x$). And (ii) is true since, for $x, y \in S$,

$$C_x \neq C_y \Rightarrow xR'y \quad \text{(by 17.3)}$$
$$\Rightarrow C_x \cap C_y = 0 \quad \text{(by 17.4)}.$$

Clarificatory note. Suppose that $x_1, x_2, x_3, \ldots$ is a complete repetition-free list of the elements of S. Then certainly

$$C_{x_1}, \quad C_{x_2}, \quad C_{x_3}, \quad \ldots \tag{*}$$

is an exhaustive list of R-classes, but (cf. 17.3) it is not in general a repetition-free list of them. Thus it would be wrong to imagine that 17.5 says that each element of S belongs to C_{x_i} for only one value of i. What 17.5 does say is that the collection of subsets that appear (whether once or repeatedly) in the list (*) is a partition of S.

For a mathematical illustration of 17.5, consider the case where S is $\mathbb{Z}$ and R is congruence modulo 5. It follows at once from 14.2(ii) that each integer belongs to precisely one of the classes C_0, C_1, C_2, C_3, C_4. So in this case a complete repetition-free list of the R-classes (forming a partition of $\mathbb{Z}$) is:

$$C_0 = \{\ldots, -10, -5, 0, 5, 10, 15, 20, \ldots\},$$
$$C_1 = \{\ldots, -9, -4, 1, 6, 11, 16, 21, \ldots\},$$
$$C_2 = \{\ldots, -8, -3, 2, 7, 12, 17, 22, \ldots\},$$
$$C_3 = \{\ldots, -7, -2, 3, 8, 13, 18, 23, \ldots\},$$
$$C_4 = \{\ldots, -6, -1, 4, 9, 14, 19, 24, \ldots\}.$$

Turning again to the general discussion of an equivalence relation R on a set S, we round off the section by noting two further simple facts.

17.6 For each $x \in S$, the one and only R-class to which x belongs is C_x (and therefore, for x, $y \in S$, $x \in C_y \Rightarrow C_x = C_y$).

Proof. Let $x \in S$. Because the set of R-classes is a partition of S, there is precisely one R-class to which x belongs; and this R-class is identified, by 17.1, as C_x.

17.7 The following are all equivalent to one another:

> (i) x, y belong to the same R-class;
> (ii) $C_x = C_y$;
> (iii) xRy.

$(x, y \in S)$

Proof. (i) and (ii) are equivalent because (by 17.6) C_x is the (unique) R-class to which x belongs and C_y is the (unique) R-class to which y belongs. And (ii) and (iii) are equivalent by 17.3. Therefore (i), (ii), (iii) are all equivalent to one another.

18. Congruence classes

Throughout this section, m will denote a fixed positive integer, and, for each integer x, C_x will denote the congruence class of $x \pmod{m}$, i.e. the R-class of x in the case where R is congruence modulo m. Thus, for each $x \in \mathbb{Z}$, we have

$$C_x = \{\ldots, x - 2m, x - m, x, x + m, x + 2m, \ldots\}.$$

From earlier sections we can obtain immediately certain simple facts about congruence classes (mod m), e.g. from 17.3, we have:

18.1 $C_x = C_y \Leftrightarrow x \equiv y \pmod{m}$ $(x, y \in \mathbb{Z})$.

By 17.5, we know that:

18.2 The collection of congruence classes (mod m) is a partition of $\mathbb{Z}$.

And thirdly, it follows from 14.2(ii) that:

18.3 There are m different congruence classes (mod m), and C_0, C_1, C_2, ..., C_{m-1} is a complete repetition-free list of them.

The set of congruence classes (mod m), i.e. $\{C_0, C_1, C_2, \ldots, C_{m-1}\}$, is denoted by $\mathbb{Z}_m$. Thus $\mathbb{Z}_m$ is a set of order m (see remark (e) of §2). While each of its member objects C_j is an infinite subset of $\mathbb{Z}$, we must be willing, when discussing $\mathbb{Z}_m$, to think of each C_j as a single object.

We can turn $\mathbb{Z}_m$ into an algebraic system by giving meanings to the sum and product of two of its member objects. This algebraic system turns out to be of very great interest, both because calculations in it provide a powerful and efficient means of gaining information about integers, and because, as will be seen in later chapters, this is just one example of many algebraic systems that can be produced by following, in a more general context, the pattern we are about to see in the case of $\mathbb{Z}_m$.

Tentatively, let us propose the following definitions of addition and multiplication of congruence classes (mod m):

$$C_x + C_y = C_{x+y} \qquad (x, y \in \mathbb{Z}) \tag{1}$$

$$C_x \cdot C_y = C_{xy} \qquad (x, y \in \mathbb{Z}). \tag{2}$$

Here a logical point of some subtlety arises because each member object of $\mathbb{Z}_m$ has many possible labels (e.g. $C_0 = C_m = C_{2m} = \ldots$), and serious attention must be given to the objection that, apparently at least, the meaning given by the equations (1), (2) to the sum and product of two congruence classes (mod m) depends on the particular labels C_x, C_y that we use for the objects being added or multiplied: if we were to use alternative labels for the same objects (say $C_{x'}$, $C_{y'}$), might we not come (via equations (1) and (2)) to a different conclusion about what the sum and product of these objects are? Happily we can, by the following short lemma, show that the attempt to define the sum and product of two congruence classes (mod m) by the equations (1), (2) survives these objections.

18.4 The equations (1) and (2) provide *consistent* definitions of $\xi + \eta$ and $\xi\eta$ (where ξ, η are members of $\mathbb{Z}_m$) in the sense that the meanings they give to $\xi + \eta$ and $\xi\eta$ are independent of the choice of labels of ξ, η.

Proof. Let ξ, η be two arbitrary member objects of $\mathbb{Z}_m$. Let C_x, $C_{x'}$ be two possible labels for ξ and C_y, $C_{y'}$ be two possible labels for η. Use of the former label in each case would lead one to the conclusions

$$\xi + \eta = C_{x+y} \quad \text{and} \quad \xi\eta = C_{xy}$$

while use of the latter label in each case would lead to

$$\xi + \eta = C_{x'+y'} \quad \text{and} \quad \xi\eta = C_{x'y'}$$

Our task, therefore, is to show that $C_{x+y} = C_{x'+y'}$ and $C_{xy} = C_{x'y'}$.

Since $C_x = C_{x'} (=\xi)$ and $C_y = C_{y'} (=\eta)$, it follows (by 18.1) that

$$x \equiv x' \,(\text{mod } m) \quad \text{and} \quad y \equiv y' \,(\text{mod } m).$$

Hence, by 14.3, $x + y \equiv x' + y' \,(\text{mod } m)$ and $xy \equiv x'y' \,(\text{mod } m)$. Therefore, by 18.1,

$$C_{x+y} = C_{x'+y'} \quad \text{and} \quad C_{xy} = C_{x'y'}$$

and our task is achieved.

It should be noted that the problem of consistency of definition will arise whenever we attempt to give a meaning to the sum, product, etc., of equivalence classes.

At this point we end §18, knowing that we have satisfactorily turned $\mathbb{Z}_m$ into an algebraic system. In §19 below, we shall consider properties of this system against the background of a very general preliminary discussion.

19. Properties of $\mathbb{Z}_m$ as an algebraic system

To begin with, suppose simply that we have a set S on which addition and multiplication are defined: i.e. for all x, $y \in S$, each of $x + y$ and xy has a meaning. Here is a list of interesting basic properties that *may* be possessed by the system $(S, +, .)$, i.e. the set S together with the addition and multiplication defined on it:

A0: $\forall x, y \in S$, $x + y$ is a member of S;

A1: (associative property of $+$) $\forall x, y, z \in S$, $(x + y) + z = x + (y + z)$;

A2: (commutative property of $+$) $\forall x, y \in S$, $x + y = y + x$;

A3: (existence of a zero) S contains a special element $\circ$ with the property that $\forall x \in S$, $x + \circ = \circ + x = x$;

A4: (existence of negatives [applicable only when A3 holds]): corresponding to each element $x \in S$, there is an element x' (also in S) such that $x + x' = x' + x = \circ$;

M0: $\forall x, y \in S$, xy is a member of S;

M1: (associative property of .) $\forall x, y, z \in S$, $(xy)z = x(yz)$;

M2: (commutative property of .) $\forall x, y \in S$, $xy = yx$;

M3: (existence of a unity) S contains a special element i (different from $\circ$ if A3 holds) with the property that $\forall x \in S$, $ix = xi = x$;

M4: (existence of inverses [applicable only when M3 holds]) corresponding to each $x \in S$, excepting $x = \circ$ if A3 holds, there is an element x^* (also in S) such that $xx^* = x^*x = i$;

D: (distributive property) $\forall x, y, z \in S$,

$$x(y + z) = xy + xz \quad \text{and} \quad (y + z)x = yx + zx.$$

When we construct or encounter for the first time a new system $(S, +, .)$, we may expect, in general, to find that it possesses some of the above properties but not others. We call the system $(S, +, .)$ a **field** iff it possesses *all* the properties in the above list. Thus a field is a system of a specially nice kind, lacking no basic property. After a little thought the student will realize that the systems of rational, real, and complex numbers are all fields.

After the above generalities, let us consider specifically the system $(\mathbb{Z}_m, +, .)$, where $m \in \mathbb{N}$.

19.1 For every integer $m \geqslant 2$, the system $(\mathbb{Z}_m, +, .)$ possesses all of the properties A0, A1, A2, A3, A4, M0, M1, M2, M3, D.

Proof. Let m be an arbitrary integer $\geqslant 2$.

From the definition of the sum and product of two congruence classes (mod m), it is clear that A0 and M0 hold for $(\mathbb{Z}_m, +, .)$.

To check A1, consider three arbitrary elements C_x, C_y, C_z of $\mathbb{Z}_m$ $(x, y, z \in \mathbb{Z})$. We have

$$
\begin{aligned}
(C_x + C_y) + C_z &= C_{x+y} + C_z && \text{(by the definition of addition on } \mathbb{Z}_m) \\
&= C_{(x+y)+z} && \text{(by the definition of addition on } \mathbb{Z}_m) \\
&= C_{x+(y+z)} && \text{(by associative property of addition of integers)} \\
&= C_x + C_{y+z} && \text{(by definition of addition on } \mathbb{Z}_m) \\
&= C_x + (C_y + C_z) && \text{(by definition of addition on } \mathbb{Z}_m).
\end{aligned}
$$

Hence A1 holds for the system $(\mathbb{Z}_m, +, .)$.

Similarly, for arbitrary elements C_x, C_y of $\mathbb{Z}_m$, we have

$$C_x + C_y = C_{x+y} = C_{y+x} = C_y + C_x$$

proving that A2 holds.

By the same kind of approach, we can show that M1, M2, D all hold too. The congruence class C_0 acts as a zero, since

$$\forall x \in \mathbb{Z}, \quad C_x + C_0 = C_0 + C_x = C_x$$

And C_1 (which is different from C_0 since $1 \not\equiv 0 \pmod{m}$, m being $\geqslant 2$) acts as a unity, since

$$\forall x \in \mathbb{Z}, \quad C_1 C_x = C_x C_1 = C_x$$

Thus A3 and M3 hold.

Finally, we note that, for every $x \in \mathbb{Z}$

$$C_x + C_{(-x)} = C_{(-x)} + C_x = C_{x+(-x)} = C_0$$

and (C_0 being the "zero") this establishes that A4 holds.

We could re-phrase 19.1 by saying that, for $m \geqslant 2$, the system $(\mathbb{Z}_m, +, .)$ is a field iff it possesses the further property M4. The next theorem gives the interesting completion of this story.

19.2 For $m \geqslant 2$,

(i) if m is composite, $(\mathbb{Z}_m, +, .)$ is not a field,

(ii) if m is prime, $(\mathbb{Z}_m, +, .)$ is a field.

Proof. (i) Suppose that m ($\geqslant 2$) is composite. Then $m = kl$ for some $k, l \in \mathbb{N}$ satisfying $1 < k, l < m$, and we have

$$C_0 = C_m = C_{kl}$$

i.e.

$$C_0 = C_k C_l \tag{α}$$

Since $1 < k < m$, $k \not\equiv 0$ (*mod* m); and therefore $C_k \neq C_0$. Hence, if $(\mathbb{Z}_m, +, .)$ were a field, there would be, in $\mathbb{Z}_m$, an inverse of C_k, i.e. an element $C_u \in \mathbb{Z}_m$ ($u \in \mathbb{Z}$) such that $C_u C_k = C_1$; and from the above equation (α), we would deduce

$$C_u C_0 = C_u(C_k C_l) = (C_u C_k)C_l = C_1 C_l$$

i.e.

$$C_0 = C_l$$

which is a contradiction since (because $1 < l < m$) $l \not\equiv 0$ (mod m).

It follows that $(\mathbb{Z}_m, +, .)$ is not a field in this case.

(ii) Suppose that m is prime.

Consider an arbitrary nonzero member C_x of $\mathbb{Z}_m$ ($x \in \mathbb{Z}$). Because $C_x \neq C_0$, $x \not\equiv 0$ (mod m) and so $m \nmid x$. Therefore (by 12.1), since m is prime, hcf$(m, x) = 1$, and hence (by 12.3) there are integers u, v such that $ux + vm = 1$. From this it follows that $ux \equiv 1$ (mod m), i.e. $C_{ux} = C_1$, i.e.

$$C_u C_x = C_x C_u = C_1 \qquad (= \text{the unity of } \mathbb{Z}_m).$$

This shows that C_u is an inverse of C_x in $\mathbb{Z}_m$.

Thus the existence of an inverse of each nonzero member of $\mathbb{Z}_m$ is established, and so $(\mathbb{Z}_m, +, .)$ is a field in this case.

It is a useful easy exercise to make up addition and multiplication tables for $\mathbb{Z}_m$ for selected values of m. As an illustration the addition and multiplication tables for $\mathbb{Z}_5$ are given below.

+	C_0	C_1	C_2	C_3	C_4		$\times$	C_0	C_1	C_2	C_3	C_4
C_0	C_0	C_1	C_2	C_3	C_4		C_0	C_0	C_0	C_0	C_0	C_0
C_1	C_1	C_2	C_3	C_4	C_0		C_1	C_0	C_1	C_2	C_3	C_4
C_2	C_2	C_3	C_4	C_0	C_1		C_2	C_0	C_2	C_4	C_1	C_3
C_3	C_3	C_4	C_0	C_1	C_2		C_3	C_0	C_3	C_1	C_4	C_2
C_4	C_4	C_0	C_1	C_2	C_3		C_4	C_0	C_4	C_3	C_2	C_1

(*Explanation*: the sum $C_x + C_y$ and the product $C_x C_y$ in $\mathbb{Z}_5$ are indicated in the row headed C_x and the column headed C_y in the left and right-hand tables, respectively.)

Were we interested primarily in number theory (i.e. properties of integers), we could now use our considerable knowledge of the systems $(\mathbb{Z}_m, +, .)$ to establish non-trivial theorems about integers, e.g. Fermat's theorem on congruences, which states that if p is a prime and n is an integer not divisible by p, then $n^{p-1} \equiv 1$ (mod p). However, it is more efficient and it gives more insight to pursue our stated aim in this textbook—the study of algebraic systems in general. In pursuing that aim, we shall obtain theorems that apply in particular to $\mathbb{Z}_m$ and enable many interesting results about integers like Fermat's theorem to be deduced in a relatively easy way.

EXERCISES ON CHAPTER THREE

1. A relation $\sim$ is defined on $\mathbb{R}$ by $x \sim y \Leftrightarrow [x] = [y]$ $(x, y \in \mathbb{R})$. Prove that $\sim$ is an equivalence relation on $\mathbb{R}$ and that, for each $n \in \mathbb{Z}$, the $\sim$-class of n is the interval $[n, n+1)$.

2. Let A, T be given sets, A being a subset of T, and let $S = \mathcal{P}(T)$. Let a relation α be defined on S by $X \alpha Y \Leftrightarrow X \cap A = Y \cap A$ $(X, Y \in S)$. (i) Prove that α is an equivalence relation. (ii) Prove that the α-class of $\emptyset$ is $\mathcal{P}(T - A)$. (iii) If $|A| = n$, where $n \in \mathbb{N}$, how many different α-classes are there?

3. Give examples of relations ρ, σ, τ on $\mathbb{Z}$ such that ρ is reflexive and symmetric but not transitive, σ is reflexive and transitive but not symmetric, and τ is symmetric and transitive but not reflexive. (The existence of these examples shows that no two of the reflexive, symmetric and transitive properties imply the third.)

4. A relation ρ is defined on $\mathbb{R}^2$ by

$$(a, b)\rho(c, d) \Leftrightarrow a + d = b + c \qquad (a, b, c, d \in \mathbb{R}).$$

Show that ρ is an equivalence relation. Interpreting $\mathbb{R}^2$ as the coordinate plane, give a geometrical description of the partition of $\mathbb{R}^2$ into ρ-classes.

5. A relation R is defined on $\mathbb{N}$ by

$$xRy \Leftrightarrow \exists n \in \mathbb{Z} \text{ s.t. } x = 2^n y \qquad (x, y \in \mathbb{N}).$$

Prove that R is an equivalence relation. Show that no R-class contains more than one prime. What is the smallest value of x $(\in \mathbb{N})$ for which the R-class of x contains no prime?

6. Let S be the set of all $n \times n$ matrices with real entries, and let a relation ρ be defined on S by

$$A \rho B \Leftrightarrow \text{there exist } r, s \in \mathbb{N} \text{ such that } A^r = B^s \qquad (A, B \in S).$$

Prove that ρ is an equivalence relation on S.

7. Draw up a multiplication table for $\mathbb{Z}_6$. How many solutions are there in $\mathbb{Z}_6$ of the equation $x^2 = x$?

8. Taking a lead from the proof of 19.2(ii), find which of the classes $C_1, C_2, \ldots, C_{96}$ in the field $\mathbb{Z}_{97}$ is the inverse of C_{41}. Find the set of all integers satisfying $41x \equiv 2 \pmod{97}$.

9. An equivalence relation β is defined on $\mathbb{Z}$ by

$$x \beta y \Leftrightarrow x^2 \equiv y^2 \pmod 5 \qquad (x, y \in \mathbb{Z}).$$

How many different β-classes are there?

Let B_t stand for the β-class of t $(t \in \mathbb{Z})$. Suppose that the rules

$$\text{(i) } B_x + B_y = B_{x+y} \qquad \text{(ii) } B_x \cdot B_y = B_{xy} \qquad (x, y \in \mathbb{Z})$$

are proposed as definitions of the sum and product of two β-classes. Investigate whether each of (i), (ii) should be accepted as a consistent definition.

CHAPTER FOUR

MAPPINGS

20. Introduction

The idea of a mapping is one of the most widely occurring in mathematics. This chapter is devoted to a survey of associated terminology, notations and theorems which will be frequently used in later chapters of the book (and are also in continual use in other parts of mathematics). Because this material plays such a key role, students will be well advised (whether or not they have any prior knowledge of mappings) to study this chapter closely until they have gained a detailed understanding of its contents.

Definition. Let S, T be non-empty sets (not necessarily different from each other). A **mapping** from S to T is a rule which associates with (or produces from) *each* element x of S a *single* element (depending on x) belonging to T.

Simple illustrations. (1) The "integral part function", which produces from each real number x its integral part, the integer $[x]$, is an example of a mapping from $\mathbb{R}$ to $\mathbb{Z}$.

(2) Let Γ be the coordinate plane. Then rotation anti-clockwise about the origin through $30°$ is a mapping from Γ to Γ: it is, in a certain sense, a rule which associates with (or produces from) each point X in Γ a single point X', also in Γ and related to X as shown in the diagram.

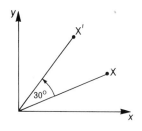

Remarks. (*a*) As the illustrations suggest, the word *rule* in the definition must be understood in a very general sense. Indeed it might be argued that the definition would convey its meaning more clearly if *rule* were replaced by a more general word or phrase such as *idea* or *mathematical notion.*

(*b*) The student should note particularly the force of the italicized words *each* and *single* in the definition.

(*c*) *Function, map, transformation* are among synonyms used for *mapping* in certain mathematical contexts.

(*d*) We usually denote mappings by single letters (such as $f, g, \ldots,$ $F, G, \ldots, \theta, \phi, \ldots$).

(*e*) In the case of a mapping f from S to T, the set S is called the **domain** of f, and the set T is called the **codomain** of f.

(*f*) The fact that f is a mapping from S to T is conveniently displayed symbolically as follows:

$$f : S \to T.$$

(*g*) Let f be a mapping from S to T. For each $x \in S$, the single element of T that f associates with (or produces from) x is denoted by $f(x)$. The correct technical term for $f(x)$ is the **image** of x under f. Another form of words often used is as follows: if $f(x) = y$, one says that "f maps x to y".

(*h*) It is sometimes helpful to picture a mapping $f : S \to T$ by means of the diagram below. The rectangular areas represent the sets S, T, and the dots within represent the elements of these sets. The "action" of f (i.e. the detail of how f associates a single element of T with each element of S) is indicated by drawing an arrow from x to $f(x)$ for each $x \in S$.

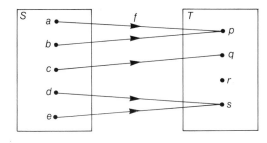

Here: $f(a) = f(b) = p$
$f(c) = q$
$f(d) = f(e) = s.$

(*i*) To specify a particular mapping $f : S \to T$, one has to say what $f(x)$ is for each element $x \in S$. Often this is most naturally achieved by means of a general formula telling (in terms of x) what $f(x)$ is for each $x \in S$, e.g. a mapping $f : \mathbb{R} \to \mathbb{R}$ may be specified by the formula

$$f(x) = x^2 + 1 \qquad (x \in \mathbb{R}).$$

A useful variant exists. One may specify a mapping from S to T by writing

$$x \longmapsto \boxed{} \quad (x \in S)$$

filling in the shaded blank with a formula for the image of x ($x^2 + 1$, or whatever it might be).

(j) For far-reaching technical reasons, it is important to be clear about what we mean when two mappings f, g are called *equal*. In fact, we call the two mappings f, g equal iff all the following conditions hold:

(i) f, g have the same domain (S, say);

(ii) f, g have the same codomain;

(iii) $\forall x \in S, f(x) = g(x)$.

(k) This final remark is addressed only to students who are already to a great extent familiar with the idea of a mapping, and it may be ignored meantime by others. It is possible to give a more formal definition of a mapping—namely that a mapping is a 3-tuple (S, T, G) in which S and T are non-empty sets (the domain and codomain of the mapping) and G is a subset of $S \times T$ with the property that for each $x \in S$ there is precisely one member of G with first object x. As with the formal definition of a relation in chapter 3, it is left to the interested student to think out why this formal definition makes sense and what advantages it offers over the informal approach.

21. The image of a subset of the domain; surjections

Let f be a mapping from S to T, and let A be a subset of the domain S.
The set

$$\{f(x) : x \in A\}$$

(i.e. the subset of the codomain T comprising all the images under f of elements of A) is called the image of the set A under f and is denoted by $f(A)$. Thus:

21.1 For $y \in T$, $y \in f(A) \Leftrightarrow \exists x \in A$ s.t. $y = f(x)$.

A case of special interest is that in which A is the whole of S. We call the subset $f(S)$ of T the **image set** of f and denote it by $\operatorname{im} f$.

Illustration. Let f be the mapping from $\{0, 1, 2, 3, 4, 5\}$ to $\{0, 1, 2, 3\}$ specified by

$$f(0) = 0, \quad f(1) = 0, \quad f(2) = 0, \quad f(3) = 1, \quad f(4) = 1, \quad f(5) = 3;$$

and let $A = \{0, 3\}$, $B = \{0, 1, 3\}$, $C = \{0, 1, 2\}$. Then, as the student should verify to his satisfaction,

$$f(A) = \{0, 1\}, \quad f(B) = \{0, 1\}, \quad f(C) = \{0\}, \quad \text{and} \quad \operatorname{im} f = \{0, 1, 3\}.$$

It will be noted that we have now introduced three different usages of the word "image" in connection with a mapping $f : S \rightarrow T$, viz. (i) the image under f of an element of S, (ii) the image under f of a subset of S, (iii) the image set of f. It is important that the student should be able to avoid confusion among these three usages.

As the above illustration shows, the image set of a mapping need not be the whole of the codomain. This observation prompts a definition.

Definition. Let $f : S \rightarrow T$ be a mapping. We say that f is a **surjection** (or that f is a surjective mapping) iff im f is the whole of the codomain T.

The following simple result is the basis of the way we usually test whether a given mapping is surjective.

21.2 Let $f : S \rightarrow T$ be a mapping. Then f is surjective iff

$$\forall y \in T, \ \exists x \in S \text{ s.t. } y = f(x)$$

Proof. The condition

$$\exists x \in S \text{ s.t. } y = f(x) \qquad (y \in T)$$

holds if $y \in \text{im} f$ and does not hold if $y \in T - \text{im} f$ (see 21.1). Therefore the condition holds for all $y \in T$ iff im f is the whole of T, i.e. iff f is surjective.

Remark. The word "onto" has traditionally been used, both as a preposition and as an adjective, to indicate surjections. For any mapping $f : S \rightarrow T$, one may say that f maps S to T. It is common to vary this by saying that f maps S onto T when f is surjective; and it is also common to speak of "an onto mapping", meaning simply "a surjection".

This section concludes with some interesting little results about images of subsets of the domain of a mapping.

As a preliminary, note that clearly:

21.3 If $f : S \rightarrow T$ is a mapping and A, B are subsets of S such that $A \subseteq B$, then $f(A) \subseteq f(B)$.

We employ this simple observation in proving the following theorem.

21.4 Let $f : S \rightarrow T$ be a mapping, and let A, B be subsets of S. Then:
  (i) $f(A \cap B) \subseteq f(A) \cap f(B)$;
  (ii) $f(A \cup B) = f(A) \cup f(B)$.

Proof. (i) Since $A \cap B \subseteq A$ and $A \cap B \subseteq B$, it follows from 21.3 that

$$f(A \cap B) \subseteq f(A) \quad \text{and} \quad f(A \cap B) \subseteq f(B).$$

Hence (by 7.2(i)) $f(A \cap B) \subseteq f(A) \cap f(B)$.

(ii) Similarly, it follows from 21.3 and 7.2(ii) that

$$f(A) \cup f(B) \subseteq f(A \cup B).$$

To establish the reverse containment, suppose that $y \in f(A \cup B)$. Then $y = f(x)$ for some $x \in A \cup B$. The element x must belong to A or to B, and so $y(=f(x))$ belongs to $f(A)$ or to $f(B)$. Therefore $y \in f(A) \cup f(B)$.
This shows that $y \in f(A \cup B) \Rightarrow y \in f(A) \cup f(B)$ $\qquad (y \in T)$
i.e. that

$$f(A \cup B) \subseteq f(A) \cup f(B).$$

The result now follows.

Post-script to 21.4(i). There are instances where $f(A \cap B)$ is not equal to $f(A) \cap f(B)$; e.g. if f is the mapping from $\mathbb{Z}$ to $\mathbb{Z}$ given by $x \mapsto x^2$ and if $A = \{0, 2\}$ and $B = \{-2, 0\}$, then, as is easily checked, $f(A) \cap f(B) = \{0, 4\}$, while $f(A \cap B) = \{0\}$.

22. Injections; bijections; inverse of a bijection

Definition. A mapping $f : S \rightarrow T$ is called an **injection** (or an injective mapping) iff in all cases distinct elements of S have distinct images under f. Thus f is injective iff

$$a \neq b \Rightarrow f(a) \neq f(b) \qquad (a, b \in S) \qquad (\alpha)$$

Illustration. Consider the mappings f_2 and f_3 from $\mathbb{R}$ to $\mathbb{R}$ given by $f_2(x) = x^2$, $f_3(x) = x^3$ ($x \in \mathbb{R}$). The mapping f_2 is not injective: for it is possible to find two distinct elements of $\mathbb{R}$ with the same image under f_2 (e.g. -5 and 5); but f_3 is injective because different real numbers have different cubes.

By applying the law of contraposition to the clause (α) in the definition, we see at once that:

22.1 The mapping $f : S \rightarrow T$ is injective iff

$$f(a) = f(b) \Rightarrow a = b \qquad (a, b \in S)$$

This result 22.1 is often used to prove that a given mapping is injective. It is also clear from the definition of an injection that:

22.2 If $f : S \rightarrow T$ is an injection and A is a finite subset of S, then $|f(A)| = |A|$.

In the remainder of this section we are concerned with mappings which are both surjective and injective.

Definition 1. A mapping $f : S \rightarrow T$ is called a **bijection** (or a bijective mapping) iff f is both surjective and injective.
And, for the special case $S = T$:

Definition 2. A bijection $f : S \to S$ is called a **permutation** of S.

[*Note*: In certain elementary contexts, people tend to use the word *permutation* to mean arrangement in a definite order of some given objects: e.g. they may describe 53124 as a "permutation of the objects 1, 2, 3, 4, 5". But, at the level of mathematical education now reached, it is very much preferable to describe 53124 as a "*rearrangement* of 12345" and to reserve the word *permutation* to mean a mapping of the kind detailed in definition 2.]

Taking up the general discussion of bijective mappings, suppose that $f : S \to T$ is a bijection. For each $y \in T$, there is exactly one element $x \in S$ satisfying $y = f(x)$. (There is at least one since f is surjective, and there cannot be more than one since f is injective.) It follows that we can specify a mapping from T to S by

$$y \mapsto x \qquad (y \in T)$$

where x is the unique answer to the question "f(what in S) $= y$?". The mapping from T to S so specified is called the **inverse** of f and is denoted by f^{-1}.

So, to re-state the essential point in a less discursive way, when f is a bijection from S to T, f^{-1} is the mapping from T to S given by

$$f^{-1}(y) = \text{the element of } S \text{ that } f \text{ maps to } y \qquad (y \in T).$$

We note and record:

22.3 If $f : S \to T$ is a bijection, then

$$f^{-1}(y) = x \Leftrightarrow f(x) = y \qquad (x \in S, y \in T)$$

Illustration. Let E be the set of even integers and let f be the (clearly bijective) mapping from $\mathbb{Z}$ to E given by $f(x) = 2x \ (x \in \mathbb{Z})$. Then f^{-1} is the mapping from E to $\mathbb{Z}$ given by $f^{-1}(y) = \frac{1}{2}y \ (y \in E)$.

A significant observation is that whenever we have a bijection (f, say) from S to T, there is set up for us a "one-to-one correspondence" between the elements of S and the elements of T (see diagram): each element x of S is put in correspondence with the single element $f(x)$ of T, and each element y of T is put in correspondence with the single element $f^{-1}(y)$ of S.

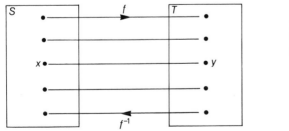

$$y = f(x)$$
and
$$x = f^{-1}(y)$$

Such thinking brings us close to the following result.

22.4 Let S, T be non-empty finite sets. Then there exists a bijection from S to T iff $|S| = |T|$.

Proof. (i) Suppose that there is a bijection f from S to T. Then

$$|S| = |f(S)| \quad \text{(by 22.2, since } f \text{ is injective)}$$
$$= |T| \quad \text{(since } f \text{ is surjective).}$$

(ii) Conversely, suppose that $|S| = |T|$ ($=n$, say). Let $x_1, x_2, \ldots, x_n$ and $y_1, y_2, \ldots, y_n$ be complete repetition-free lists of the elements of S and T, respectively. Then the mapping $f : S \to T$ given by

$$f(x_i) = y_i \quad (i = 1, 2, \ldots, n)$$

is clearly a bijection from S to T.

Proof of the stated result is now complete.

23. Restriction of a mapping

Consider the following mappings f and g:

$$f : \mathbb{R} \to \mathbb{R} \quad \text{given by} \quad f(x) = x^2;$$
$$g : [0, \infty) \to [0, \infty) \quad \text{given by} \quad g(x) = x^2.$$

Although one might say that f and g have the same action (which could loosely be described as "squaring"), f and g are not equal (e.g. f, g do not have the same domain; moreover g^{-1} exists whereas f^{-1} does not). As the next definition makes clear, the relationship between these similar but unequal mappings is expressed by saying that g is a restriction of f.

Definition. Let $f : S \to T$ be a mapping, and let A, B be non-empty subsets of S, T, respectively, satisfying $f(A) \subseteq B$. Then we can specify a mapping $g : A \to B$ by

$$g(x) = f(x) \quad (x \in A)$$

This mapping g is called the **restriction** of f to A as domain and B as codomain; and any mapping produced from f in this way is called a restriction of f.

When a non-bijective mapping is given, it is often possible and profitable to consider a bijective restriction of it. Given any mapping f, we can produce a surjective restriction of f just by changing the codomain to $\text{im} f$; and often a simple change of domain then produces a bijective restriction of f.

To illustrate what has just been said, consider the sine function, i.e. the mapping from $\mathbb{R}$ to $\mathbb{R}$ given by $x \mapsto \sin x$. This mapping is neither surjective nor injective. However, a surjective restriction is obtained on reducing the codomain to $\text{im}(\sin)$, i.e. to $[-1, 1]$; and then a bijective restriction is produced by reducing the domain to $[-\frac{1}{2}\pi, \frac{1}{2}\pi]$. (Consideration of a sketch of the curve $y = \sin x$ will make all this clear.) The bijective restriction described—a mapping from $[-\frac{1}{2}\pi, \frac{1}{2}\pi]$ to $[-1, 1]$—is important in calculus because its inverse is used there as the inverse sine function (denoted by $\sin^{-1}$ or arcsin).

Often one is concerned with a restriction of $f : S \rightarrow T$ to a smaller domain $A\ (\subset S)$, but with the same codomain T as f itself. That restriction is denoted by $f|A$ and is usually called simply "the restriction of f to A".

24. Composition of mappings

Definition. Let $f : S \rightarrow T$ and $g : T \rightarrow U$ be mappings with the property that the domain of g is the same set as the codomain of f. Then a mapping from S to U may be specified by

$$x \mapsto g(f(x)) \qquad (x \in S)$$

This mapping is denoted by $g \circ f$ and is called the **composition** of f followed by g. (The diagram below may help with the assimilation of this definition.)

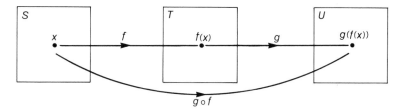

Note that (i) $g \circ f$ is not defined unless the domain of g equals the codomain of f, and (ii) when this condition is satisfied, i.e. when $g \circ f$ does exist,

domain of $g \circ f =$ domain of f,

codomain of $g \circ f =$ codomain of g.

Worked example. Let f, g be the mappings from $\mathbb{R}$ to $\mathbb{R}$ given by

$$f(x) = x^2, \quad g(x) = 2x + 1 \qquad (x \in \mathbb{R})$$

In this case both $g \circ f$ and $f \circ g$ exist. Find formulae specifying them.

Solution. For $x \in \mathbb{R}$, we have:

$$\begin{aligned}
(g \circ f)(x) &= g(f(x)) \qquad &\text{(by definition of } g \circ f) \\
&= g(x^2) \qquad &\text{(by definition of } f) \\
&= 2x^2 + 1 \qquad &\text{(by definition of } g).
\end{aligned}$$

Thus $g \circ f$ is the mapping from $\mathbb{R}$ to $\mathbb{R}$ given by $x \mapsto 2x^2 + 1$ $(x \in \mathbb{R})$. Similarly, for $x \in \mathbb{R}$,

$$(f \circ g)(x) = f(g(x)) = f(2x + 1) = (2x + 1)^2,$$

so that $f \circ g$ is the mapping from $\mathbb{R}$ to $\mathbb{R}$ given by $x \mapsto (2x + 1)^2$ $(x \in \mathbb{R})$.

In the above example, $(g \circ f)(1) = 3$ while $(f \circ g)(1) = 9$, which shows that

$g \circ f \neq f \circ g$. Thus, even in a case where $g \circ f$ and $f \circ g$ both exist and have the same domain and codomain, $g \circ f$ and $f \circ g$ may be unequal. We express this by saying that "mapping composition is not commutative".

In contrast, a very important positive property of mapping composition (called the "associative property") is that, for mappings f, g, h,

$$(f \circ g) \circ h = f \circ (g \circ h)$$

provided that the composite mappings involved exist. Here now is a formal statement of it.

24.1 Let $h : S \to T$, $g : T \to U$, $f : U \to V$ be mappings. Then

$$(f \circ g) \circ h = f \circ (g \circ h)$$

Proof. We note first that $f \circ g$ exists and is a mapping from T to V. Hence $(f \circ g) \circ h$ exists and is a mapping from S to V. Similarly, we find that $f \circ (g \circ h)$ exists and is a mapping from S to V. Thus $(f \circ g) \circ h$ and $f \circ (g \circ h)$ are mappings with the same domain and codomain. Further, for every $x \in S$,

$$
\begin{aligned}
[(f \circ g) \circ h](x) &= (f \circ g)[h(x)] &&\text{(by meaning of } k \circ h) \\
&= f\{g[h(x)]\} &&\text{(by meaning of } f \circ g) \\
&= f\{(g \circ h)(x)\} &&\text{(by meaning of } g \circ h) \\
&= [f \circ (g \circ h)](x) &&\text{(by meaning of } f \circ l).
\end{aligned}
$$

It follows that $(f \circ g) \circ h = f \circ (g \circ h)$.

Identity mappings
For each non-empty set S, a mapping denoted by i_S from S to S is specified by the simple rule $x \mapsto x$ $(x \in S)$, i.e.

$$i_S(x) = x \qquad (x \in S)$$

This mapping i_S is called the **identity mapping** from S to itself.

It is easily seen that:

24.2 For every set S, i_S is a permutation of S.

An important property of identity mappings is that they behave in mapping composition like the number 1 in multiplication. That is:

24.3 Let $f : S \to T$ be a mapping. Then

$$\text{(i) } f \circ i_S = f, \quad \text{and} \quad \text{(ii) } i_T \circ f = f.$$

Proof. (i) $f \circ i_S$ and f are both mappings from S to T. Further, for every $x \in S$,

$$
\begin{aligned}
(f \circ i_S)(x) &= f(i_S(x)) &&\text{(by meaning of } f \circ g) \\
&= f(x) &&\text{(by definition of } i_S).
\end{aligned}
$$

It follows that $f \circ i_S = f$.

Part (ii) is proved similarly.

The final result in this section is a formal expression of the fact that, when f is a bijection, f and f^{-1} are mappings which undo each other's effects (e.g. in an illustration in §22, the action of f was to double while the action of f^{-1} was to halve.)

24.4 Let $f : S \rightarrow T$ be a bijection, so that f^{-1} exists. Then
$$\text{(i) } f \circ f^{-1} = i_T \quad \text{and} \quad \text{(ii) } f^{-1} \circ f = i_S.$$

Proof. (i) It is easily checked that $f \circ f^{-1}$ exists and is a mapping from T to T. So $f \circ f^{-1}$ and i_T are mappings with the same domain and codomain.

Let y be an arbitrary element of T, and let $f^{-1}(y) = x$. Then (see 22.3) $y = f(x)$, and hence we have
$$(f \circ f^{-1})(y) = f(f^{-1}(y)) = f(x) = y = i_T(y).$$

It follows that $f \circ f^{-1} = i_T$.

Part (ii) can be proved in a fairly similar way, and so its proof is left as an exercise.

25. Some further results and examples on mappings

In 24.4 we showed that if f is a bijection from S to T and $g = f^{-1}$, then
$$g \circ f = i_S \quad \text{and} \quad f \circ g = i_T \tag{1}$$

We now prove (as a kind of converse) that the existence of a mapping g satisfying (1) implies that f is bijective and that a mapping g satisfying (1) can only be f^{-1}. More precisely:

25.1 Let $f : S \rightarrow T$ be a mapping. Suppose that a mapping $g : T \rightarrow S$ exists and is such that
$$g \circ f = i_S \quad \text{and} \quad f \circ g = i_T$$
Then f is bijective, and g is its inverse.

Proof. f is injective because (cf. 22.1), for $a, b \in S$,
$$f(a) = f(b) \Rightarrow g(f(a)) = g(f(b))$$
$$\Rightarrow (g \circ f)(a) = (g \circ f)(b)$$
$$\Rightarrow i_S(a) = i_S(b) \qquad (\text{since } g \circ f = i_S)$$
$$\Rightarrow a = b.$$

Moreover, f is surjective: for (cf. 21.2) if y is an arbitrary element of T, y is the image under f of an element of S, viz. $g(y)$, since
$$f(g(y)) = (f \circ g)(y) = y \qquad (f \circ g \text{ being equal to } i_T).$$

It follows that f is bijective, and therefore f^{-1} exists. Further:

$$f^{-1} = f^{-1} \circ i_T \quad \text{(by 24.3)}$$
$$= f^{-1} \circ (f \circ g) \quad \text{(since } f \circ g = i_T)$$
$$= (f^{-1} \circ f) \circ g \quad \text{(by 24.1)}$$
$$= i_S \circ g \quad \text{(by 24.4)}$$
$$= g \quad \text{(by 24.3)}.$$

The result is now completely proved.

The next two results are corollaries of 25.1.

25.2 Let $f : S \to T$ and $g : T \to U$ be bijections. Then $g \circ f$ is also a bijection, and its inverse equals $f^{-1} \circ g^{-1}$.

Proof. It is easily checked that $(g \circ f) \circ (f^{-1} \circ g^{-1})$ exists. Moreover:

$$(g \circ f) \circ (f^{-1} \circ g^{-1}) = g \circ [f \circ (f^{-1} \circ g^{-1})]$$
$$= g \circ [(f \circ f^{-1}) \circ g^{-1}] \quad \text{(by 24.1)}$$
$$= g \circ (i_T \circ g^{-1}) = g \circ g^{-1} = i_U \quad \text{(by 24.3 and 24.4)}.$$

Similarly, $(f^{-1} \circ g^{-1}) \circ (g \circ f)$ exists and equals i_S. So the equations

$$h \circ (g \circ f) = i_S, \qquad (g \circ f) \circ h = i_U$$

are satisfied with $h = f^{-1} \circ g^{-1}$ (a mapping from U to S).
By 25.1, the stated result follows immediately.
An even easier application of 25.1 yields:

25.3 If f is a bijection, then f^{-1} is also a bijection and $(f^{-1})^{-1} = f$.

Particular cases of 25.2 and 25.3 give us:

25.4 (i) If f, g are permutations of the set S, then so is $g \circ f$.
(ii) If f is a permutation of the set S, then so is f^{-1}.

Worked examples
1. Let $f : S \to T$, $g : T \to U$, $h : T \to U$ be mappings such that (i) $g \circ f = h \circ f$ and (ii) f is surjective. Prove that $g = h$.

Solution. Let t be an arbitrary element of T. Since f is surjective, $t = f(x)$ for some $x \in S$. Hence we have:

$$g(t) = g(f(x))$$
$$= (g \circ f)(x)$$
$$= (h \circ f)(x) \quad \text{(by (i))}$$
$$= h(f(x))$$
$$= h(t)$$

c

Since g and h have the same domain and codomain, it follows that $g = h$.

2. Let $f : S \to T, g : T \to U$ be mappings such that $g \circ f$ is injective. Prove that f is injective.

Solution. For $a, b \in S$,

$$
\begin{aligned}
f(a) = f(b) &\Rightarrow g(f(a)) = g(f(b)) \\
&\Rightarrow (g \circ f)(a) = (g \circ f)(b) \\
&\Rightarrow a = b \quad \text{(since $g \circ f$ is injective)}.
\end{aligned}
$$

Hence f is injective.

EXERCISES ON CHAPTER FOUR

1. Each of the following attempts to specify a mapping contains a howler. Spot the howler (and resolve never to imitate it).

(i) A mapping $f : \mathbb{N} \to \mathbb{N}$ is specified by $f(x) = x - 7$ $(x \in \mathbb{N})$.

(ii) A mapping $g : \mathbb{R} \to \mathbb{R}$ is specified by $g(x) = 1/x$ $(x \in \mathbb{R})$.

(iii) A mapping $h : \mathbb{R} \to \mathbb{R}$ is specified by $h(x) = \begin{cases} x + 1 & \text{if } x \geqslant 0, \\ 0 & \text{if } x \leqslant 0. \end{cases}$

(iv) A mapping $\theta : \mathbb{Q} \to \mathbb{Z}$ is specified by $\theta(m/n) = m + n$ $(m, n \in \mathbb{Z}, n \neq 0)$.

2. Look carefully at the definition of equality of mappings (§20, remark (j)) and decide whether, in the context of mappings from the coordinate plane Γ to itself:

(i) rotation anticlockwise about O through $360° = i_\Gamma$;

(ii) rotation anticlockwise about O through $180° =$ rotation clockwise about O through $180°$.

3. The mapping $f : \mathbb{R} \to \mathbb{R}$ is given by $f(x) = x^4 - 1$ $(x \in \mathbb{R})$.

(i) What is the image of 2 under f?

(ii) What is the image under f of the set $[-3, 2]$?

(iii) What is im f? Is f surjective?

4. Let S, T be finite non-empty sets with $|S| = m$, $|T| = n$.

(i) How many different mappings are there from S to T?

(ii) How many different injections are there from S to T (a) if $m > n$, (b) if $m \leqslant n$?

(iii) Show that if $m = n + 1$, the number of different surjections from S to T is $\frac{1}{2} n\{(n+1)!\}$. Find a corresponding expression for the number of surjections from S to T in the case $m = n + 2$.

5. Let f, g be the mappings from $\mathbb{N}$ to $\mathbb{N}$ given by

$$f(x) = 2x, \quad g(x) = [\tfrac{1}{2}(x+1)] \quad (x \in \mathbb{N}).$$

Prove that f is injective but not surjective, and that g is surjective but not injective.

Give examples of (a) a mapping from $\mathbb{N}$ to $\mathbb{N}$ that has infinite image set but is neither surjective nor injective, (b) a bijection from $\mathbb{N}$ to $\mathbb{Z}$.

6. Suppose that $f : S \to T$ is a surjection. Prove that, for all $A \subseteq S$, $T - f(A) \subseteq f(S - A)$.

7. Suppose that $f : S \to T$ is injective. Prove that, for $A, B \subseteq S$,

$$f(A \cap B) = f(A) \cap f(B).$$

8. Let $f : S \to T$ be a mapping. Prove that

$$f \text{ is injective} \Leftrightarrow \forall A \subseteq S, f(A) \cap f(S - A) = \emptyset.$$

9. Let $f : S \to T$ be a mapping, let A be a subset of S, and let

$$B = \{x \in S : f(x) \in f(A)\}.$$

(i) Show that $A \subseteq B$.

(ii) Prove that $A = B$ if f is injective.

(iii) Determine B in the particular case where $S = T = \mathbb{R}$, $A = [-1, 3]$, and f is given by $x \mapsto x^2$.

10. The relation ρ is defined on $\mathbb{N} \times \mathbb{N}$ by

$$(a, b)\rho(c, d) \Leftrightarrow a + d = b + c$$

Prove that ρ is an equivalence relation on $\mathbb{N} \times \mathbb{N}$.

For $(a, b) \in \mathbb{N} \times \mathbb{N}$, let $C_{(a, b)}$ denote the ρ-class of (a, b); and let S denote the set of all

ρ-classes. Show that the formula

$$\theta(C_{(a,b)}) = b - a \qquad (a, b \in \mathbb{N})$$

consistently defines a mapping from S to $\mathbb{Z}$ and that this mapping θ is bijective.

11. Let $f : S \to T$ be a mapping. Let the relation R be defined on S by

$$xRy \Leftrightarrow f(x) = f(y) \qquad (x, y \in S).$$

Show that R is an equivalence relation on S.

(i) If S, T are finite, what can be said about the number of different R-classes (a) if f is injective, (b) if f is surjective?

(ii) In the general case, let U be the set of all R-classes. Prove that f can be expressed in the form $\alpha \circ \beta$, where α is an injection from U to T and β is a surjection from S to U.

12. Let S be an arbitrary non-empty set. Prove that there does not exist a bijection from S to $\mathscr{P}(S)$. [*Hint*: Suppose $f : S \to \mathscr{P}(S)$ is a bijection and consider $T = \{x \in S : x \notin f(x)\}$.]

13. Let $f : \mathbb{R} \to \mathbb{R}$ be defined by $f(x) = x^2 - 4x + 5 \quad (x \in \mathbb{R})$.

(i) Draw the curve $y = f(x)$.

(ii) What is the image set B of f?

(iii) What number c is such that f has bijective restrictions $f_1 : (-\infty, c] \to B$, $f_2 : [c, \infty) \to B$?

(iv) Find formulae for $f_1^{-1}(y)$ and $f_2^{-1}(y) \quad (y \in B)$.

14. Suppose that $f : S \to T$ and $g : T \to U$ are mappings such that $g \circ f$ is surjective. Prove that g is surjective.

15. Suppose that $f : S \to T$, $g : S \to T$, $h : T \to U$ are mappings such that $h \circ f = h \circ g$ and h is injective. Prove that $f = g$.

16. Mappings f, g, h from the set S to itself are such that $f \circ g = h \circ f = i_S$. Prove that f is bijective and that $f^{-1} = g = h$.

17. Suppose that the mapping $f : S \to T$ is not surjective. Let $U = \{0, 1\}$. Prove that there exist two unequal mappings g_1, g_2 from T to U such that $g_1 \circ f = g_2 \circ f$.

18. Suppose that $f : S \to T$ is injective (but not necessarily surjective). Prove the existence of a mapping $g : T \to S$ such that $g \circ f = i_S$.

CHAPTER FIVE

SEMIGROUPS

26. Introduction

At this point in the book we launch into the mainstream of abstract algebra, i.e. into the general study of algebraic systems.

When we refer to an *algebraic system*, we always mean a non-empty set whose elements can be combined by addition, or by multiplication, or by both, or by some other such binary operation(s). In this chapter we begin with a formal look at binary operations, and then we set the scene for abstract algebra by discussing the very simple kind of algebraic system known as a *semigroup*. This discussion enables us to introduce (in a context not cluttered by extraneous details) basic concepts such as *identity elements*, *inverses*, *subsystems*—concepts that are relevant to all the species of algebraic system studied in this book. In chapters 6 to 8 we shall consider in detail the important special kind of semigroup that we call a *group*. In a semigroup (in particular in a group) there is just one binary operation. That is why it seems sensible to study semigroups and groups before rings and fields, systems with two binary operations (addition and multiplication) which we explore in the final chapter of the book.

27. Binary operations

Throughout this section S will denote a set.

We say that $\circ$ is a **binary operation** on S iff $x \circ y$ has a meaning for every x, $y \in S$. In this context, we think of $\circ$ not merely as the symbol that is put on paper between x and y to form $x \circ y$ but as the process or rule (call it what you will) by which $x \circ y$ is formed from x and y (see also remark (e) below).

In the case $S = \mathbb{Z}$, familiar examples of binary operations include $+$ (addition), $-$ (subtraction), . (multiplication).

Remarks. (a) We use symbols like $\circ$, $*$ as general symbols for binary operations. Often, however, we deal with binary operations with special names (e.g. addition, subtraction, multiplication) and then we follow familiar conventions in notation for these binary operations (e.g. $+$, $-$, . or $\times$ for addition, subtraction, multiplication, respectively). Moreover, when we deal with a binary operation called multiplication, the simple notation xy (instead of $x.y$ or $x \times y$) is usually acceptable for the "product" of two elements x, y.

(b) If $\circ$ is a binary operation on S, it must not be presupposed that $x \circ y$ and $y \circ x$ are equal for all x, $y \in S$.

(c) Suppose that $\circ$ is a binary operation on S (so that $x \circ y$ has a meaning for all x, $y \in S$). Then, for any subset T of S, it is obviously true that $x \circ y$ has a meaning for all x, $y \in T$. Thus it is legitimate to claim that $\circ$ is a binary operation on each subset T of S.

(d) Let $\circ$ be a binary operation on S. This we have defined to mean simply that $x \circ y$ has a meaning for all x, $y \in S$. Unlike some authors, we do *not* insist that in all cases the object $x \circ y$ must be a member of S. Our point of view has the advantage of giving to the following definition the prominence it merits.

Definition. $\circ$ being a binary operation on S, we say that S is **closed** under $\circ$ iff $x \circ y$ is a member of S for all x, $y \in S$.

In the light of remark (c), this means that, more generally, a subset T of S is closed under $\circ$ iff $x \circ y \in T$ for all x, $y \in T$ (e.g. $\mathbb{Z}$ is closed under $-$ because, for every x, $y \in \mathbb{Z}$, $x - y$ is again an integer. But the subset $\mathbb{N}$ is not closed under $-$ since, for example, although $2, 4 \in \mathbb{N}$, $2 - 4 \notin \mathbb{N}$.)

(e) Some textbooks define a binary operation to be a mapping. The point is that if $\circ$ is a binary operation (in the sense detailed above) and if, for all x, $y \in S$, $x \circ y$ is an element of the set A, then one may regard $\circ$ as a mapping from $S \times S$ to A, viz. the mapping given by

$$(x, y) \mapsto x \circ y$$

This point of view will not be adopted here, because persistence with it makes certain considerations of subsets of S unnecessarily difficult.

28. Associativity and commutativity

We now think about two basic properties that a binary operation may possess.

Definition 1. Suppose that $\circ$ is a binary operation on the set S and that S is closed under $\circ$.   We say that $\circ$ is **associative** (on S) iff

$$\forall x, y, z \in S, \quad (x \circ y) \circ z = x \circ (y \circ z)$$

Most binary operations encountered in elementary mathematics are associative. An obvious exception among binary operations on $\mathbb{R}$ is $-$ (subtraction): although $\mathbb{R}$ is closed under $-$, $-$ is non-associative, since, for example,

$$(1 - 2) - 3 = -4, \text{ while } 1 - (2 - 3) = 2.$$

Definition 2. If ∘ is a binary operation on S, we say that ∘ is **commutative** (on S) iff $\forall x, y \in S, x \circ y = y \circ x$.

For example, addition and multiplication are commutative on $\mathbb{R}$, but subtraction is not commutative on $\mathbb{R}$ since, for example, $1 - 2 \neq 2 - 1$.

More generally, if ∘ is a binary operation on S and $x, y \in S$, we say that x, y **commute** with each other (under ∘) iff $x \circ y = y \circ x$. So ∘ is commutative on S iff every pair of elements of S commute with each other (under ∘).

For the rest of this section, suppose that ∘ is an associative binary operation on the set S (S being closed under ∘).

Because ∘ is associative, for any $x, y, z \in S$, we can write

$$x \circ y \circ z \quad \text{(without brackets!)}$$

without risk of ambiguity: for, by the definition of associativity, the two possible interpretations of $x \circ y \circ z$ (viz. $(x \circ y) \circ z$ and $x \circ (y \circ z)$) are the same.

Likewise, if w, x, y, z all belong to S, then, by repeated application of the definition of associativity, we can show that the five ways of inserting brackets into "$w \circ x \circ y \circ z$" yield identical results:

$$\begin{aligned}
\{w \circ (x \circ y)\} \circ z &= \{(w \circ x) \circ y\} \circ z \\
&= (w \circ x) \circ (y \circ z) \quad \text{(since } (a \circ y) \circ z = a \circ (y \circ z)) \\
&= w \circ \{x \circ (y \circ z)\} \quad \text{(since } (w \circ x) \circ b = w \circ (x \circ b)) \\
&= w \circ \{(x \circ y) \circ z\}
\end{aligned}$$

More generally (∘ being associative on S), every expression

$$x_1 \circ x_2 \circ x_3 \circ \ldots \circ x_n \quad \text{(where } x_1, x_2, x_3, \ldots, x_n \in S),$$

of whatever length, is unambiguous in the sense that all ways of bracketing it yield the same answer. This can be proved formally by induction on n.

If ∘ is commutative as well as associative on S, it is not difficult to see that, in the expression

$$x_1 \circ x_2 \circ x_3 \circ \ldots \circ x_n \quad \text{(where each } x_i \in S)$$

one can make any prescribed change in the order of the x_i's without altering the meaning of the expression.

29. Semigroups: definition and examples

Definition. Let S be a non-empty set on which is defined a binary operation $\ast$. Then the system $(S, \ast)$ [i.e. the algebraic system consisting of the set S and the binary operation $\ast$ on S] is called a **semigroup** iff

(1) S is closed under $\ast$ and (2) $\ast$ is associative on S.

For example, clearly $(\mathbb{Z}, +)$ is a semigroup. Other examples will be given shortly.

The definition of a semigroup, as given above, has been worded so as to stress, for the beginner's benefit, that a semigroup is not merely a set of objects but a set endowed with a binary operation (satisfying certain conditions). Nevertheless it is customary to use phraseology that hides this point, e.g. as an alternative to the sentence "$(S, *)$ is a semigroup", one frequently says "S is a semigroup with respect to $*$" (or even simply "S is a semigroup", when the binary operation can be surmised from the context); and it is commonplace to speak of "the semigroup S" instead of "the semigroup $(S, *)$". A corresponding remark applies to every kind of algebraic system that we study.

Notice that, in the definition of a semigroup, we do not insist that the binary operation is commutative. A semigroup $(S, *)$ is called a **commutative semigroup** iff $*$ is commutative on S.

A *multiplicative* semigroup means a semigroup in which the binary operation is multiplication. Similarly, an *additive* semigroup means one in which the binary operation is addition.

Examples of semigroups

(1) $\mathbb{Z}$ is a semigroup with respect to multiplication. So is $\mathbb{N}$.

(2) Let Ω be a non-empty set, and let S be the set of all mappings from Ω to Ω. Then $(S, \circ)$ is a semigroup, where $\circ$ denotes mapping composition.

(3) Let $S = \mathscr{P}(\Omega)$, where Ω is a given non-empty set. Then $(S, \cap)$ and $(S, \cup)$ are commutative semigroups. [Note that $(S, \cap)$ and $(S, \cup)$ are different semigroups: for, although they involve the same set S, they involve different binary operations.]

(4) The set of all $n \times n$ real matrices (n being a given positive integer) is a multiplicative semigroup.

Proof of the above assertions is a trivial matter. In each case, we need only remark that the set is non-empty and closed under the binary operation, and that the binary operation is associative.

30. Powers of an element in a semigroup

Let $(S, *)$ be a semigroup, and let $x \in S$. Because of the associativity of $*$, we can unambiguously define (for each $n \in \mathbb{N}$) the nth **power** of x under $*$ by

$$n\text{th power of } x = x * x * x * \ldots * x \qquad (n \text{ factors}).$$

If $*$ is multiplication, the notation x^n is appropriate. If $*$ is addition, the notation nx is appropriate. [nx, meaning

$$x + x + x + \ldots + x \qquad (n \text{ terms}),$$

is usually called the nth *multiple* of x.]

The following *index laws* can be proved for every semigroup.

30.1 For $m, n \in \mathbb{N}$ and x an element of the semigroup $(S, *)$,

   (i) (mth power of x) $*$ (nth power of x) = $(m+n)$th power of x,

   (ii) nth power of (mth power of x) = (mn)th power of x.

It is instructive to write down the forms taken by these laws (a) in a multiplicative semigroup, (b) in an additive semigroup. These are

   (a) (i) $x^m . x^n = x^{m+n}$,   (ii) $(x^m)^n = x^{mn}$;

   (b) (i) $mx + nx = (m+n)x$,   (ii) $n(mx) = (mn)x$.

Proof of 30.1(i). Left-hand side $= \underbrace{(x * x * \ldots * x)}_{m \text{ factors}} * \underbrace{(x * x * \ldots * x)}_{n \text{ factors}}$

$$= \underbrace{x * x * x * x * \ldots * x}_{(m+n) \text{ factors}}$$

$$= (m+n)\text{th power of } x.$$

Part (ii) may be proved similarly.

31. Identity elements and inverses

Definition. In a semigroup $(S, *)$ an element e belonging to S is called an **identity** iff $e * x = x * e = x$ for every $x \in S$.

It should be appreciated that a given semigroup may or may not have an identity, e.g. in the semigroup $(\mathbb{Z}, .)$, the element 1 is an identity; but the semigroup $(E, .)$, where E is the set of all even integers, has no identity element.

31.1 A semigroup cannot contain more than one identity element.

Proof. Suppose that $(S, *)$ is a semigroup and that the elements e, f of S are both identities. [We shall prove that $e = f$.] We have:

$$e = e * f \quad \text{(since } f \text{ is an identity)}$$
$$= f \quad \text{(since } e \text{ is an identity)}.$$

The result follows.

Notice that the proof of 31.1 does not use associativity.

When we know that a semigroup S has an identity e, we are, by 31.1, justified in calling e *the* identity of S.

In an additive semigroup S, it is usual to call the identity (if there is one) *zero* and denote it by 0: it satisfies

$$x + 0 = 0 + x = x \text{ for all } x \in S.$$

Likewise, in a multiplicative semigroup S, the identity element (if there is one) is often denoted by 1: it satisfies

$$1x = x1 = x \text{ for all } x \in S.$$

We turn attention now to inverses.

Definition. Let $(S, *)$ be a semigroup with an identity element e, and let $x \in S$. An element y belonging to S is called an **inverse** of x (in the semigroup) iff $x * y = y * x = e$.

If $(S, *)$ is a semigroup with identity e, then (since $e * e = e$) the element e has itself as an inverse, but it would be wrong to think that every other element of S must necessarily have an inverse: each may or may not. For example, in the semigroup $(\mathbb{Z}, .)$, which has identity 1, the elements 1 and -1 have themselves as inverses, but no other element has an inverse in the semigroup. [Of course one cannot, for example, claim $\frac{1}{2}$ as an inverse of 2 in this system: for an inverse of 2 must be an element of the semigroup, i.e. of $\mathbb{Z}$.]

We now establish that inverses, when they exist, are unique.

31.2 Let $(S, *)$ be a semigroup with an identity element, and let x be an element of S. Then x cannot have more than one inverse in S.

Proof. Suppose that y, z are both inverses of x in S. Then

$$y = y * e \qquad \text{(where } e \text{ is the identity of the semigroup)}$$
$$ = y * x * z \qquad \text{(since } z \text{ is an inverse of } x \text{ [note also that associativity is tacitly used here])}$$
$$ = e * z \qquad \text{(since } y \text{ is an inverse of } x \text{)}$$
$$ = z \qquad (e \text{ being the identity)}.$$

This proves the result.

Because of 31.2, where a semigroup element x has an inverse, it is justifiable to refer to that inverse as *the* inverse of x.

In a multiplicative semigroup with identity 1, the natural notation for the inverse of element x (if it exists) is x^{-1}: it satisfies $xx^{-1} = x^{-1}x = 1$.

In an additive semigroup with an identity (i.e. a zero, 0), the inverse of element x (if it exists) is naturally called the *negative* of x and denoted by $-x$: it satisfies $x + (-x) = (-x) + x = 0$.

32. Subsemigroups

Suppose that $(S, *)$ is a semigroup and that T is a subset of S. (Then $*$ is a binary operation on T, as explained in §27, remark (c).) We say that T is a **subsemigroup of** S iff the system $(T, *)$ is a semigroup.

For example, in the semigroup $(\mathbb{Z}, .)$, $\mathbb{N}$ is a subsemigroup (since $\mathbb{N}$ is a subset of $\mathbb{Z}$ and $(\mathbb{N}, .)$ is a semigroup).

Clearly, for any semigroup S, S is a subsemigroup of itself.

The following theorem shows how simply one can test whether a given non-empty subset of a semigroup is a subsemigroup.

32.1 Suppose that T is a non-empty subset of S, where $(S, *)$ is a semigroup. Then T is a subsemigroup of S iff T is closed under $*$.

Proof. Obviously, if T is a subsemigroup of S, so that $(T, *)$ is a semigroup, then T is closed under $*$.

Conversely, suppose that T is closed under $*$. Then, since the condition

$$(x * y) * z = x * (y * z)$$

holds for all values of x, y, z in S, it holds, in particular, for all values of x, y, z in T, and thus $*$ is associative on T. From the definition of a semigroup, it now follows that $(T, *)$ is a semigroup, i.e. that T is a subsemigroup of S.

The stated result is now fully proved.

Worked example. Let S be the multiplicative semigroup of all 2×2 real matrices, and let T be the subset of S consisting of the matrices of the form $\begin{bmatrix} x & 0 \\ 0 & 0 \end{bmatrix}$ $(x \in \mathbb{R})$. Prove that T is a subsemigroup of S.

Solution. Clearly $T \neq \emptyset$.

Consider two arbitrary elements $A = \begin{bmatrix} x & 0 \\ 0 & 0 \end{bmatrix}$, $B = \begin{bmatrix} y & 0 \\ 0 & 0 \end{bmatrix}$ of T $(x, y \in \mathbb{R})$.

The product $AB = \begin{bmatrix} xy & 0 \\ 0 & 0 \end{bmatrix}$, which also belongs to T.

Hence T is closed under multiplication.

It follows (by 32.1) that T is a subsemigroup of S.

Post-script to this example. The semigroup S has an identity, viz. the matrix $I = \begin{bmatrix} 1 & 0 \\ 0 & 1 \end{bmatrix}$. And, although $I \notin T$, the subsemigroup T has an identity, viz. the matrix $\begin{bmatrix} 1 & 0 \\ 0 & 0 \end{bmatrix}$. This last assertion is clearly true since

$$\begin{bmatrix} 1 & 0 \\ 0 & 0 \end{bmatrix}\begin{bmatrix} x & 0 \\ 0 & 0 \end{bmatrix} = \begin{bmatrix} x & 0 \\ 0 & 0 \end{bmatrix}\begin{bmatrix} 1 & 0 \\ 0 & 0 \end{bmatrix} = \begin{bmatrix} x & 0 \\ 0 & 0 \end{bmatrix} \quad \text{for all } x \in \mathbb{R}.$$

The point is one which may surprise the uninitiated:

32.2 In a semigroup S with an identity, the identity (if it exists) of a subsemigroup is not necessarily the same as that of S.

EXERCISES ON CHAPTER FIVE

1. A binary operation $\circ$ is defined on $\mathbb{Z}$ by

$$x \circ y = x + y - xy \qquad (x, y \in \mathbb{Z}).$$

Prove that $(\mathbb{Z}, \circ)$ is a semigroup. Let $T = \{x \in \mathbb{Z} : x \leqslant 1\}$. Prove that T is a subsemigroup of the semigroup $(\mathbb{Z}, \circ)$.

2. Let $S = \mathscr{P}(\Omega)$, where Ω is a given set. Do the semigroups $(S, \cap)$ and $(S, \cup)$ have identities? If so, what are they?

3. In a semigroup $(S, *)$ an element $e \in S$ is called a *left identity* iff $e * x = x$ for all $x \in S$; an element $f \in S$ is called a *right identity* iff $x * f = x$ for all $x \in S$.

(i) Let S be a non-empty set, and let an operation $\circ$ be defined on S by $x \circ y = y$ $(x, y \in S)$. Prove that $(S, \circ)$ is a semigroup in which every element is a left identity.

(ii) Prove that, if in a semigroup there is more than one left identity, then there is no right identity.

4. In a semigroup $(S, *)$ an element x is called *idempotent* iff $x * x = x$.

Let S be a finite multiplicative semigroup, and let $a \in S$. Prove the existence of positive integers k, l such that $a^k = a^{k+l}$, and prove further that a^{kl} is idempotent. (This shows that, for every element a in a finite semigroup, there is a power of a which is idempotent.)

5. Suppose that a multiplicative semigroup S contains a left identity e (see question 3) and that correspondingly every element of S has a left inverse (i.e. for each $x \in S$, there is an element $x' \in S$ satisfying $x'x = e$ but not necessarily $xx' = e$). Prove that, for every $x \in S$, xx' is idempotent (see question 4). Deduce that e is an identity (i.e. a right identity as well as a left identity) and that, for each element x, its left inverse x' is in fact an inverse.

CHAPTER SIX

AN INTRODUCTION TO GROUPS

33. The definition of a group

A **group** may be concisely defined as a semigroup with an identity, in which every element has an inverse.

By 31.1 and 31.2, we can say straight away that:

33.1 In any group, there is just one identity element, and each element of the group has a unique inverse.

A very familiar example of a group is provided by the set of nonzero real numbers (i.e. $\mathbb{R} - \{0\}$): this set, it is easily seen, is a group with respect to multiplication.

Further examples of groups are given in §34. Meanwhile, in the remaining paragraphs of this section, there is a short sequence of remarks whose natural place is immediately beside the definition of a group. The most basic properties of groups will be developed in §35, and later in the chapter attention is given to subgroups (§§36, 37), to the very important concept of the period of a group element (§38), and to cyclic groups (§39), these being groups of a particularly simple kind.

Remarks. (*a*) The binary operation in a group need not be commutative. In the special case when the binary operation is commutative, the group is called an **abelian group**. (The adjective derives from the name of the nineteenth-century mathematician N. H. Abel.)

(*b*) For convenience, we shall concentrate in our general discussion of groups almost exclusively on multiplicative groups, i.e. groups in which the binary operation is multiplication. So, except where otherwise stated, the sentence "*G* is a group" should be taken to mean that *G* is a multiplicative group, i.e. that $(G, .)$ is a group. Correspondingly, in general discussion, the identity element of a group will be denoted by 1, and the inverse of an element x by x^{-1}. It is assumed that, if and when the need arises, the student will be capable of translating definitions and results (given here for a multiplicative group) into the form appropriate for an additive group (or, more generally, a group in which the binary operation is $*$). Indeed it is recommended that, in the early stages of his study of groups, the student should systematically practise this capability. As an example, consider the "reversal rule for inverses" (to be proved in §35), which, for a multiplicative group G, states:

$$\forall x, y \in G, \quad (xy)^{-1} = y^{-1}x^{-1}$$

Fairly obviously, the corresponding result for an additive group G is:

$$\forall x, y \in G, \quad -(x+y) = (-y)+(-x)$$

which can be re-stated

$$\forall x, y \in G, \quad -(x+y) = -y-x$$

since in an additive group one defines subtraction by $u-v = u+(-v)$.

(c) Besides the concise definition of a group with which we opened this section, the following fuller (but obviously equivalent) definition of a multiplicative group should be learned.

A set $G(\neq \emptyset)$ on which multiplication is defined is a (multiplicative) group iff all the following postulates are satisfied:

G0: G is closed under multiplication;

G1: multiplication is associative on G;

G2: there is an element $1 \in G$ such that $1x = x1 = x$ for all $x = G$;

G3: corresponding to each element x of G, there is an element $x^{-1} \in G$ such that $x^{-1}x = xx^{-1} = 1$.

The postulates G0, G1, G2, G3 are called the *group postulates* (or **group axioms**). When given a set G on which multiplication is defined, we can prove that G is a group by showing that G0, G1, G2, G3 all hold for the given set G. (Notice that it is not necessary to check separately that $G \neq \emptyset$, since $G \neq \emptyset$ is implied by G2.)

34. Examples of groups

The following examples illustrate the very widespread occurrence of groups in mathematics. Proofs that the systems mentioned are groups are in most cases omitted.

(1) *Examples of groups from the familiar number systems*
$\mathbb{Z}, \mathbb{Q}, \mathbb{R}, \mathbb{C}$ are all abelian groups with respect to addition.

$\mathbb{Q}-\{0\}, \mathbb{R}-\{0\}, \mathbb{C}-\{0\}$ are all abelian groups with respect to multiplication.

(2) *From matrix algebra*

(i) For given $m, n \in \mathbb{N}$, the set of all $m \times n$ matrices with entries in $\mathbb{R}$ is an abelian group with respect to matrix addition. The same is true with $\mathbb{R}$ replaced by $\mathbb{C}$.

(ii) For given $n \in \mathbb{N}$, the set of all $n \times n$ nonsingular matrices with entries in $\mathbb{R}$ is a group with respect to matrix multiplication. (I_n is the identity; each matrix X in the set has an inverse (the matrix X^{-1}) also in the set.) This group is non-abelian if $n \geqslant 2$. It is often denoted by $GL(n, \mathbb{R})$, GL standing for "general linear". If instead we had considered the set of all $n \times n$ matrices with entries in $\mathbb{C}$, we would have described the group $GL(n, \mathbb{C})$.

(3) *From number theory*

For each integer $m \geqslant 2$, $\mathbb{Z}_m$ (see §§18, 19) is a group with respect to addition of congruence classes.

For each prime number p, $\mathbb{Z}_p - \{C_0\}$, i.e. the set of nonzero congruence classes modulo p, is a group with respect to multiplication of congruence classes.

(4) *Groups of permutations*

Let Ω be a non-empty set, and let G be the set of all permutations of Ω (i.e. bijections from Ω to itself). Then G is a group with respect to mapping composition. We call $(G, \circ)$ the **symmetric group** on Ω.

[*Proof* that $(G, \circ)$ is a group. We show that the group axioms all hold for G (taking mapping composition as the "multiplication").

G0: Holds, by 25.4(i).

G1: Holds, since mapping composition is associative (24.1).

G2: Holds since the mapping i_Ω belongs to G (by 24.2) and satisfies (by 24.3) $f \circ i_\Omega = i_\Omega \circ f = f$ for all $f \in G$.

G3: Holds because, for each $f \in G$, the inverse mapping f^{-1} exists, belongs to G (by 25.4(ii)), and (by 24.4) satisfies $f \circ f^{-1} = f^{-1} \circ f = i_\Omega$.]

A number of other matters deserve mention under the current heading of "groups of permutations".

(*a*) The phrase "a group of permutations of the set Ω" means a group whose elements are some (not necessarily all) of the permutations of Ω, the binary operation being mapping composition. Those who pursue their study of group theory beyond the level of this textbook will discover that groups of permutations play a most interesting key role in the development of the general theory of groups.

(*b*) In the context of group theory, when discussing a group G of permutations (of some set Ω), we often abbreviate $f \circ g$ to fg ($f, g \in G$).

(*c*) *The symmetric group of degree n* (where $n \in \mathbb{N}$) means the group of all $n!$ permutations of a set of n objects, which, for convenience, we label $1, 2, 3, \ldots, n$. This group is denoted by S_n.

An element of S_n is often denoted by a $2 \times n$ matrix in which the top row has entries $1, 2, 3, \ldots, n$ (in order) and the second row gives (in order) the images of $1, 2, 3, \ldots, n$, respectively, under the element in question; e.g. the permutation in S_4 that maps 1 to 3, 2 to 2, 3 to 4, and 4 to 1 may be denoted by

$$\begin{pmatrix} 1 & 2 & 3 & 4 \\ 3 & 2 & 4 & 1 \end{pmatrix}$$

(5) *Groups of symmetries of a geometrical figure*

Suppose that Π is a geometrical figure in a euclidean space S. (S might be a plane or ordinary 3-dimensional space.) We think of S as a set of points.

An **isometry** of S means a bijective mapping from S to S that preserves distance; i.e. two points at distance d from each other must have images under the mapping which have distance d from each other. Examples include translations, rotations, reflections.

An isometry of S is called a **symmetry** of Π iff it leaves Π as a whole unchanged while perhaps permuting its component elements (points, lines, etc.); e.g. if S is a plane and Π is a square in the plane with centre O, rotation of the plane about O through a right angle (in either direction) is a symmetry of Π.

It can be proved that the set of symmetries of a given figure Π is, in all cases, a group with respect to mapping composition. We call this group the **symmetry group** of Π. In it the identity is the mapping i_S, which fixes (i.e. maps to itself) every point of the space S.

As a specific example, consider the symmetry group of the equilateral triangle Δ with centroid at O and altitudes along the lines l, m, n (see diagram).

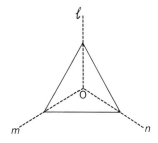

Clearly Δ has the following symmetries:

     $1 =$ identity mapping from the plane to itself,
     $p =$ rotation anticlockwise about O through $120°$,
     $q =$ rotation clockwise about O through $120°$,
     $a =$ reflection in l,
     $b =$ reflection in m,
     $c =$ reflection in n.

In fact, these six symmetries form the complete group G of symmetries of the plane figure Δ.

A multiplication table for this group G is exhibited below. In the table, the entry in the row headed x and the column headed y gives the product xy. (As in the discussion of a group of permutations, so here, we use xy as an abbreviation for $x \circ y$.)

	1	p	q	a	b	c
1	1	p	q	a	b	c
p	p	q	1	c	a	b
q	q	1	p	b	c	a
a	a	b	c	1	p	q
b	b	c	a	q	1	p
c	c	a	b	p	q	1

To illustrate how the information in the table can be obtained, let us consider pa (which means the composition of a followed by p). By means of the diagram below we can follow how the vertices of the triangle are moved when we apply first a and then p, and we observe that the net result is the same as the effect of c: hence we deduce that $pa = c$.

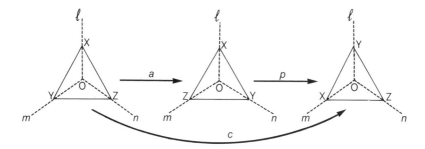

Remarks. (a) Since (as the table reveals) $ap \neq pa$, this group G is non-abelian. Later we shall prove that there is no smaller non-abelian group, i.e. that every group with fewer than 6 elements is abelian.

(b) The above table illustrates a fact that we shall prove in the next section—that, in each row or column of the multiplication table of a finite group, each element of the group appears precisely once.

(c) This subsection has brought to light a point of some practical significance—that wherever there is geometrical symmetry (e.g. in the structure of a molecule or in the configuration of atoms in a crystal), there will be a group to discuss. Often details of the symmetry and details of the group are inter-related in an intelligible way.

(6) More finite groups

If G is a finite group, $|G|$ (i.e. the number of elements in the set G) is always called the **order** of the group G; e.g. the order of S_n is $n!$ (see subsection (4) above), and the order of the group of symmetries of an equilateral triangle is 6.

Here now are details of two further interesting finite groups.

(i) The set of complex numbers $\{1, -1, i, -i\}$ is, with respect to multiplication, a group of order 4. The student will have no difficulty in constructing the multiplication table of this abelian group.

(ii) The set of matrices $Q = \{I, -I, J, -J, K, -K, L, -L\}$, where

$$I = \begin{bmatrix} 1 & 0 \\ 0 & 1 \end{bmatrix}, \quad J = \begin{bmatrix} i & 0 \\ 0 & -i \end{bmatrix}, \quad K = \begin{bmatrix} 0 & 1 \\ -1 & 0 \end{bmatrix}, \quad L = \begin{bmatrix} 0 & i \\ i & 0 \end{bmatrix}$$

is a group with respect to matrix multiplication, I being the identity. This noteworthy non-abelian group of order 8 is called the **quaternion group**. After verifying that

$$JK = -KJ = L, \quad KL = -LK = J, \quad LJ = -JL = K, \quad J^2 = K^2 = L^2 = -I.$$

one can quickly construct a multiplication table for Q. It will be worth while for the student to do so and store the result for future reference.

Trivial groups

A group of order 1 is described as a **trivial** group. In such a group the sole element is the identity, 1, and the full story of the multiplication is told in the equation $1.1 = 1$.

35. Elementary consequences of the group axioms

Throughout this section G will denote an arbitrary multiplicative group.

35.1 (The *cancellation laws* for group G) For $a, x, y \in G$,

    (i) $ax = ay \Rightarrow x = y$,

    (ii) $xa = ya \Rightarrow x = y$.

Proof of (i). Suppose that $ax = ay \, (a, x, y \in G)$. By G3, a has an inverse a^{-1} in G, and hence we have

$$a^{-1}ax = a^{-1}ay \quad \text{(associativity (G1) being tacitly used here),}$$

i.e. $\qquad\qquad\qquad\qquad\qquad\qquad 1x = 1y,$

i.e. $\qquad\qquad\qquad\qquad\qquad\qquad x = y.$

This proves (i). Part (ii) may be proved similarly.

It should be obvious why 35.1 is entitled "cancellation laws". The next result is sometimes referred to as "the *division laws* (for group G)".

35.2 For $a, b, x \in G$,

(i) $ax = b \Leftrightarrow x = a^{-1}b$,

(ii) $xa = b \Leftrightarrow x = ba^{-1}$.

Proof of (i). $ax = b \Rightarrow a^{-1}ax = a^{-1}b$ (using G3 and G1)

$$\Rightarrow 1x = a^{-1}b$$
$$\Rightarrow x = a^{-1}b$$

and, conversely,

$$x = a^{-1}b \Rightarrow ax = aa^{-1}b \Rightarrow ax = 1b \Rightarrow ax = b.$$

Part (ii) is proved similarly.

These "division laws" prompt the remark that, in a non-abelian group G, it would be a great mistake to employ the notation $\dfrac{b}{a}$ for the solution of either of the equations $ax = b$, $xa = b$: for the notation $\dfrac{b}{a}$ is ambiguous—one does not know whether it means ba^{-1} or $a^{-1}b$.

Hitherto, for y, u belonging to the group G, in order to be able to draw the conclusion "u is the inverse of y", we have had to establish *both* that $uy = 1$ *and* that $yu = 1$. However, as a particular case of 35.2, we now have:

35.3 For u, y belonging to the group G, if we know *either* that $uy = 1$ *or* that $yu = 1$, then we may conclude that $u = y^{-1}$.

Proof. By 35.2(i), $yu = 1 \Rightarrow u = y^{-1}1 = y^{-1}$;

and, by 35.2(ii), $uy = 1 \Rightarrow u = 1y^{-1} = y^{-1}$.

This last result enables all of the following to be rapidly proved.

35.4 In the group G, the inverse of the identity is itself (i.e. $1^{-1} = 1$).

35.5 For every x in the group G, $(x^{-1})^{-1} = x$.

35.6 For all x, y in the group G, $(xy)^{-1} = y^{-1}x^{-1}$.

The result 35.6 is known as the *reversal rule* for group inverses. It extends to longer products, i.e.

35.7 For elements $x_1, x_2, \ldots, x_n$ belonging to the group G,

$$(x_1 x_2 \ldots x_{n-1} x_n)^{-1} = x_n^{-1} x_{n-1}^{-1} \ldots x_2^{-1} x_1^{-1}.$$

To illustrate the usefulness of 35.3, here is a proof of 35.6.

Proof. Let $x, y \in G$. Then $y^{-1}x^{-1}$ is an element of G, and we have

$$(xy)(y^{-1}x^{-1}) = xyy^{-1}x^{-1} = x1x^{-1} = 1$$

By 35.3, it follows that $(xy)^{-1} = y^{-1}x^{-1}$.

As an important postscript to the above results, we introduce, for each element a of the group G, two mappings from G to G which we shall always denote by λ_a and ρ_a and which we call "left multiplication by a" and "right multiplication by a", respectively. These are specified by

$$\lambda_a(x) = ax, \quad \rho_a(x) = xa \qquad (x \in G).$$

It is useful to consider these mappings, largely because:

35.8 For each a belonging to the group G, λ_a and ρ_a are permutations of G, i.e. bijections from G to G.

Proof. Consider λ_a where a is an arbitrary element of G.
λ_a is injective because (cf. 22.1)

$$\lambda_a(x) = \lambda_a(y) \Rightarrow ax = ay \Rightarrow x = y \qquad \text{(by 35.1)}.$$

And λ_a is surjective because (cf. 21.2) the arbitrary element y of G is the image under λ_a of $a^{-1}y$, since $\lambda_a(a^{-1}y) = aa^{-1}y = 1y = y$.

It follows that, for each $a \in G$, λ_a is bijective.

Similarly one can prove that ρ_a is bijective for each $a \in G$.

We can now prove a fact mentioned in passing in §34, viz:

35.9 If G is a finite group, then in every row or column of the multiplication table of G each element of G appears exactly once.

Proof. Let G be a finite group of order n, and let $x_1, x_2, \ldots, x_n$ be a list of the elements of G in the order in which they appear in the headings of the multiplication table. Consider an arbitrary row of the table—say the row headed a ($a \in G$). The entries in this row of the table are, respectively,

$$ax_1, ax_2, \ldots, ax_n$$

i.e.

$$\lambda_a(x_1), \lambda_a(x_2), \ldots, \lambda_a(x_n).$$

So, since λ_a is a permutation of $G = \{x_1, x_2, \ldots, x_n\}$, it is clear that each element of G appears exactly once in the row. This proves the result for rows; and a similar discussion proves it for columns.

Powers of a group element
Let x be an element of the group G. By virtue of our discussion of positive integral powers of an element of a semigroup (in §30), x^n already has a meaning for every $n \in \mathbb{N}$. We can also define non-positive integral powers of the group element x as follows:

zeroth power: $x^0 = 1$;

negative powers: for $n \in \mathbb{N}$, $x^{-n} = (x^n)^{-1}$, which (by 35.7) is the same as $(x^{-1})^n$.

The following *index laws* are universally true in group theory.

35.10 For a group element x,

$$\text{(i) } x^m . x^n = x^{m+n} \quad \text{(ii) } (x^m)^n = x^{mn} \qquad (m, n \in \mathbb{Z})$$

To prove these, one needs to separate out various cases and patiently establish the laws in each case, making use of 30.1. By way of illustration, here is a proof of (i) for the case where $m > 0$, $n < 0$ and $|m| > |n|$.

Proof. Write $n = -l$, so that $l \in \mathbb{N}$ and $l < m$. Then

$$
\begin{aligned}
x^m . x^n &= x^m . x^{-l} \\
&= x^{m-l} . x^l . x^{-l} \quad &&\text{(using 30.1 and the fact that } m - l > 0) \\
&= x^{m-l} . x^l . (x^l)^{-1} \quad &&\text{(by definition of a negative power)} \\
&= x^{m-l} . 1 \quad &&\text{(since } yy^{-1} = 1 \text{ for every } y \text{ in the group)} \\
&= x^{m-l} \\
&= x^{m+n}
\end{aligned}
$$

36. Subgroups

Suppose that G is a (multiplicative) group, i.e. that $(G, .)$ is a group, and that H is a subset of G. (As explained in §27, remark (c), multiplication is defined on H.) We say that H is a **subgroup** of G iff $(H, .)$ is a group.

Simple illustrations:
(1) $\mathbb{Q} - \{0\}$ is a subgroup of the multiplicative group $\mathbb{R} - \{0\}$.
(2) $\{1, -1, i, -i\}$ is a subgroup of the multiplicative group $\mathbb{C} - \{0\}$.
(3) If H is a group of permutations of the set Ω and G is the symmetric group on Ω (see §34(4)), then H is a subgroup of G.

It is clear that:

36.1 For every group G, $\{1\}$ and G are subgroups of G.

We use the following terminology for subgroups of a group G:
(a) $\{1\}$ is the **trivial subgroup** of G;
(b) a **proper subgroup** of G is a subgroup different from G;
(c) a *non-trivial proper subgroup*, therefore, is a subgroup equal to neither $\{1\}$ nor G.

Two elementary facts about a subgroup are presented in the next theorem. Anyone tempted to suggest that it is unnecessary to remark on these points should scrutinize his opinions in the light of 32.2.

36.2 Suppose that H is a subgroup of the group G. Then:
(i) the identity 1_H of the group H is the same as the identity 1 of the group G;

(ii) for each $h \in H$, the inverse of h in the group H is the same as the inverse of h in the group G.

Proof. (i) Since 1_H is the identity of H, $1_H 1_H = 1_H$, and, since 1 is the identity of G, $1 1_H = 1_H$. Hence

$$1_H 1_H = 1 1_H$$

and hence, by cancellation in the group G (35.1(ii)), $1_H = 1$.

(ii) Let h be an arbitrary element of H. Let h' denote its inverse in the group H, and let h^{-1} denote its inverse in the group G. In view of part (i), $hh' = 1$. By 35.3, (since $h' \in G$) it follows that $h' = h^{-1}$. This proves the result.

Tests for a subgroup
If H is a subset of the group G, one can test whether H is a subgroup of G by checking whether the group axioms G0, G1, G2, G3 hold for H. The following results, however, give quicker and more efficient tests.

36.3 Let H be a subset of the group G. Then H is a subgroup of G iff the following conditions all hold:

(1) $H \neq \emptyset$,
(2) H is closed under multiplication,
(3) $x \in H \Rightarrow x^{-1} \in H$.

[*Note*: Condition (3) is sometimes expressed as "H is closed under the taking of inverses".]

Proof. (i) Suppose that H is a subgroup. Then (1) and (2) hold by virtue of the group axioms G2, G0 as they apply to H. Moreover (3) holds because, if $x \in H$, x has an inverse in H (since G3 holds for H) and (by 36.2(ii)) this inverse equals x^{-1}, so that $x^{-1} \in H$.

(ii) Conversely, suppose that the conditions (1), (2), (3) all hold. We prove that the group axioms G0, G1, G2, G3 hold for H.

G0 holds for H, by condition (2).

G1 clearly holds for H, since the multiplication on G is associative.

To prove G2, introduce into the discussion an element h of H (a legitimate step, by condition (1)). By (3), $h^{-1} \in H$; and hence, by (2), $hh^{-1} \in H$, i.e. $1 \in H$. Hence G2 holds for H.

G3 holds for H, by condition (3).

It follows that H is a subgroup.

Proof of the stated result is now complete.

An even simpler but not always more useful test is:

36.4 Let H be a subset of the group G. Then H is a subgroup of G iff (1) $H \neq \emptyset$ and (2) x and y belong to $H \Rightarrow x^{-1}y \in H$.

Proof. (i) If H is a subgroup, then certainly $H \neq \emptyset$ and

$$x \text{ and } y \in H \Rightarrow x^{-1} \text{ and } y \in H \qquad \text{(by G3 for } H)$$
$$\Rightarrow x^{-1}y \in H \qquad \text{(by G0 for } H).$$

(ii) Conversely, suppose that the conditions (1), (2) hold.

By (1), we may introduce an element h of H. By (2), $h^{-1}h \in H$, i.e. $1 \in H$. Therefore G2 holds for H.

Next, knowing that $1 \in H$, we deduce from (2) that

$$x \in H \Rightarrow x^{-1}1 \in H,$$

i.e.

$$x \in H \Rightarrow x^{-1} \in H.$$

Therefore G3 holds for H.

We now observe that

$$x \text{ and } y \in H \Rightarrow x^{-1} \text{ and } y \in H \qquad \text{(by the previous paragraph)}$$
$$\Rightarrow (x^{-1})^{-1}y \in H \qquad \text{(by (2))}$$
$$\Rightarrow xy \in H.$$

Therefore G0 holds for H.

Finally, G1 certainly holds for H, since the multiplication on G is associative. It follows that H is a subgroup.

Proof of the stated result is now complete.

The final variation on this theme relates to the case H finite:

36.5 Suppose that H is a *finite* subset of the group G. Then H is a subgroup of G iff (1) $H \neq \emptyset$ and (2) H is closed under multiplication.

Proof. (i) If H is a subgroup of G, then clearly the conditions (1) and (2) hold.

(ii) Conversely, suppose that the conditions (1), (2) hold.

Suppose that x is an element of H. Since H is closed under multiplication, all elements in the list $x, x^2, x^3, x^4, \ldots$ belong to H. But H is finite. Therefore the list cannot be repetition-free, i.e. we must have

$$x^r = x^s \qquad (\alpha)$$

for some $r, s \in \mathbb{N}$ with $r < s$. The equation (α) can be re-written $x^r.1 = x^r.x^{s-r}$; and hence, by cancellation in the group G, we deduce that

$$x^{s-r} = 1 \qquad (\beta)$$

Since $s-r \in \mathbb{N}$, it follows that $1 \in H$. Further, (β) can be re-written $x.x^{s-r-1} = 1$, showing (by 35.3) that $x^{-1} = x^{s-r-1}$. Since $s-r-1$ is either a positive integer or zero, i.e. x^{s-r-1} is either a positive power of x or 1, it follows that $x^{-1} \in H$.

This shows that $x \in H \Rightarrow x^{-1} \in H$. And from this and the conditions (1) and (2), it follows (by 36.3) that H is a subgroup.

Proof of the stated result is now complete.

Worked example. Let $G = GL(n, \mathbb{R})$ [see § 34(2)], and let H be the subset of G consisting of the matrices in G with determinant 1. Show that H is a subgroup of G.

[*Note.* The following facts about determinants will be assumed:
   (i) for $n \times n$ matrices A, B, $\det(AB) = \det A \times \det B$; and (consequently)
   (ii) for a nonsingular matrix A, $\det(A^{-1}) = 1/\det A$.]

Solution. We use 36.4.

(1) $H \neq \emptyset$, since the identity matrix I_n belongs to H.

(2) Let X, Y be arbitrary members of H, so that $\det X = \det Y = 1$. Then, using the facts about determinants noted above, we have

$$\det(X^{-1}Y) = \det(X^{-1}) \times \det Y = (1/\det X) \times \det Y = (1/1) \times 1 = 1;$$

and so $X^{-1}Y \in H$.

This proves that $X, Y \in H \Rightarrow X^{-1}Y \in H$.

(By 36.4) it follows that H is a subgroup of G.

Further illustrations of the use of 36.3 and 36.4 are provided by the proofs of the following theorem and of theorems in § 37.

36.6 Let H and K be subgroups of a group G. Then $H \cap K$ is also a subgroup of G.

Proof. (1) Since H, K are subgroups of G, $1 \in H$ and $1 \in K$. Hence $1 \in H \cap K$. In particular, $H \cap K \neq \emptyset$.

(2) Let x, y be arbitrary elements of $H \cap K$. Then x, y both belong to H, and so, since H is a subgroup, $x^{-1}y \in H$. Similarly, $x^{-1}y \in K$. Hence $x^{-1}y \in H \cap K$.

Thus $x, y \in H \cap K \Rightarrow x^{-1}y \in H \cap K$.

By 36.4, it follows that $H \cap K$ is a subgroup of G.

More generally, one can prove that:

36.7 If G is a group, then the intersection of any finite or infinite family of subgroups of G is again a subgroup of G.

37. Some important general examples of subgroups

(1) *Centralizers*

Let G be a given group and let a be an element of G (a to be fixed throughout the following discussion). Let

$$C(a) = \{x \in G : xa = ax\}$$

i.e. the set of all members of G that commute with the particular element a. We call $C(a)$ the **centralizer** of a (in G).

37.1 $C(a)$ is a subgroup of G.

Proof. We use 36.3.

(1) Since $1a = a1$, $1 \in C(a)$. So $C(a) \neq \emptyset$.

(2) Let $x, y \in C(a)$. Then $xa = ax$ and $ya = ay$; and hence

$$(xy)a = x(ya) = x(ay) = (xa)y = (ax)y = a(xy),$$

which shows that $xy \in C(a)$.

Thus $C(a)$ is closed under multiplication.

(3) Let $x \in C(a)$. Then $ax = xa$. Hence $x^{-1}(ax)x^{-1} = x^{-1}(xa)x^{-1}$, i.e. $x^{-1}a = ax^{-1}$, i.e. $x^{-1} \in C(a)$.

Thus $x \in C(a) \Rightarrow x^{-1} \in C(a)$.

It follows (by 36.3) that $C(a)$ is a subgroup.

(2) The centre of a group

Definition. The **centre** of a group G is the subset

$$\{x \in G : \forall g \in G, xg = gx\}$$

i.e. it is the subset of G comprising those elements that commute with every element of G. The centre of G is usually denoted by $Z(G)$.

On comparing the definition of $Z(G)$ with that of the centralizer of an element of G, we see that, for $x \in G$, x belongs to $Z(G)$ iff x belongs to the centralizer of every individual element of G. Therefore $Z(G)$ equals the intersection of all the centralizers of individual elements of G. By 36.7 and 37.1, it follows that:

37.2 For every group G, $Z(G)$ is a subgroup of G.

The result 37.2 may also be proved directly by using 36.3 instead of proceeding via 36.7 and 37.1.

In chapter 8, an illustration will be given of how useful consideration of the centre of a group can be. In the meantime, note that:

37.3 For a group G, $Z(G)$ is the whole of G iff G is abelian.

(3) The subgroup generated by an individual element

Let G be a group, and let a be an element of G (a to be fixed throughout the following discussion). Noting that (as a simple consequence of the group axioms) a^n is an element of G for every $n \in \mathbb{Z}$, consider

$$H = \{a^n : n \in \mathbb{Z}\}$$

i.e. the subset of G comprising all elements expressible as powers of a.

37.4 $H\ (=\{a^n:n\in\mathbb{Z}\})$ is a subgroup of G.

Proof. (1) Clearly $H \neq \emptyset$: e.g. $a\in H$.
  (2) Let $x, y\in H$. Then $x = a^m$, $y = a^n$ for some $m, n\in\mathbb{Z}$. Hence (by 35.10)
$$x^{-1}y = (a^m)^{-1}a^n = a^{-m}a^n = a^{n-m}$$
which shows that $x^{-1}y\in H$.
  Thus $x, y\in H \Rightarrow x^{-1}y\in H$.
  It follows (by 36.4) that H is a subgroup.

We call the subgroup $H\ (=\{a^n:n\in\mathbb{Z}\})$ the **subgroup of G generated by** a, and we denote it by $\langle a\rangle$. Note that:

37.5 If G is a group and $a\in G$, then $\langle a\rangle$ is the smallest subgroup of G containing the element a, i.e. if L is any subgroup of G such that $a\in L$, then $\langle a\rangle \subseteq L$.

Proof. Let L be a subgroup of the group G such that $a\in L$. By the group axioms as they apply to L, it is clear that every integral power of an element of L must be an element of L. In particular, therefore, $a^n\in L$ for every $n\in\mathbb{Z}$, and thus $\langle a\rangle \subseteq L$. This proves the stated result.

38. Period of an element

Let a be a given element of the group G. We are concerned throughout this section with the powers of a, i.e. the elements of the subgroup $\langle a\rangle$. We begin with a twofold definition that draws attention to an important distinction between two possible cases.

Definition. (1) If there is no positive power of a that equals 1, we say that a has **infinite period**.
  (2) If there is a positive power of a that equals 1 and the kth power of a is the lowest such positive power (i.e. $a^k = 1$ but $a^n \neq 1$ if $1 \leqslant n \leqslant k-1$), we say that a has **period** k.

Remarks. (*a*) In definition (2), we know that there is such a thing as the "lowest such positive power" by virtue of the well-ordering principle (10.1).
  (*b*) The word "order" is often used instead of "period".

Illustrations. (i) In the group $\mathbb{R}-\{0\}$, 2 has infinite period (because, $\forall n \in \mathbb{N}$, $2^n \neq 1$).
  (ii) In the group $\mathbb{C} - \{0\}$, i has period 4 (because $i^4 = 1$ while i^1, i^2, i^3 are all different from 1).
  Clearly:

38.1 In every group, the identity (and only the identity) has period 1.

There follows a detailed analysis of each of the two cases distinguished in the above definitions.

(1) *Infinite period case*

38.2 Suppose that the group element a has infinite period. Then, for $m, n \in \mathbb{Z}$, $a^m \neq a^n$ if $m \neq n$ (and consequently $\langle a \rangle$ is infinite).

Proof. For $m, n \in \mathbb{Z}$,

$$a^m = a^n \Rightarrow a^m (a^n)^{-1} = 1 \text{ and } a^n (a^m)^{-1} = 1$$
$$\Rightarrow a^{m-n} = a^{n-m} = 1$$
$$\Rightarrow a^{|m-n|} = 1$$
$$\Rightarrow |m-n| = 0 \qquad \text{(since no positive power of } a \text{ equals 1)}$$
$$\Rightarrow m = n$$

The result follows (by the law of contraposition).

An immediate corollary of 38.2 is:

38.3 In a finite group G, no element can have infinite period.

(2) *Finite period case*

38.4 Suppose that the group element a has finite period k ($k \in \mathbb{N}$). Then:
   (i) if the integer n has principal remainder r on division by k, then $a^n = a^r$;
   (ii) for $n \in \mathbb{Z}$, $a^n = 1 \Leftrightarrow k|n$;
   (iii) $1, a, a^2, \ldots, a^{k-1}$ is a complete repetition-free list of the elements of $\langle a \rangle$;
   (iv) $|\langle a \rangle| = k$, i.e. the order of the subgroup generated by a equals the period of a.

Proof. (i) Let $n \in \mathbb{Z}$ and let n have principal remainder r on division by k. Then $n = qk + r$ for some integer q. Hence, using the index laws and the fact that $a^k = 1$, we have

$$a^n = a^{qk+r} = (a^k)^q a^r = 1^q a^r = a^r$$

as required.

(ii) Let n denote an integer.

First suppose that $a^n = 1$. Let r be the principal remainder on division of n by k. Then, by part (i), $a^r = a^n = 1$. But $0 \leqslant r \leqslant k-1$ and, since k is the period of a, $a^s \neq 1$ if $1 \leqslant s \leqslant k-1$. It follows that $r = 0$, i.e. that $k|n$.

Thus $a^n = 1 \Rightarrow k|n$.

Part (ii) follows on observing that, conversely:

$$k|n \Rightarrow n = lk \text{ for some integer } l$$
$$\Rightarrow a^n = a^{lk} = (a^k)^l = 1^l = 1$$

(iii) By (i), every power of a is equal to one appearing in the list

1, a, a^2, ..., a^{k-1}. Moreover, this list is repetition-free: for if it contained a repetition, we would have $a^m = a^n$ with $0 \leqslant m < n \leqslant k-1$ and hence $a^{n-m} = 1$ with $1 \leqslant n-m \leqslant k-1$—a contradiction since the period of a is k. These observations establish part (iii).

(iv) It follows immediately from (iii) that $|\langle a \rangle| = k$.

Among all the important detail of the various parts of 38.4, the student should not overlook the simple fact that, when the group element a has period k, the sequence of successive powers of a (i.e. the sequence $\ldots, a^{-2}, a^{-1}, a^0, a, a^2, a^3, a^4, \ldots$) repeats itself with periodicity k. The case $a = i$ in the group $\mathbb{C} - \{0\}$ provides a helpful illustration of this and of all the points brought out in 38.4.

For further illustrations of group elements with finite period, let us turn to the group of all isometries of the coordinate plane. In this group, let r be rotation anticlockwise about the origin through $2\pi/n$ (i.e. through one nth of a revolution), n being an arbitrary positive integer. Clearly r has period n, and so $\langle r \rangle$ is (by 38.4(iv)) a group of order n. This illustration shows that:

38.5 For every $n \in \mathbb{N}$, there exists a group of order n.

39. Cyclic groups

Definition. A group (or subgroup) G is called **cyclic** iff there is an element a ($\in G$) such that $\langle a \rangle$ is the whole of G.

So in any group G, the subgroup generated by any element a (i.e. $\langle a \rangle$) is a cyclic subgroup of G. Thus it is easy to provide illustrations of cyclic groups, e.g.:

(1) the subgroup $\langle 2 \rangle = \{2^n : n \in \mathbb{Z}\}$ of $\mathbb{R} - \{0\}$ is an infinite cyclic group;

(2) the subgroup $\langle i \rangle = \{1, i, i^2, i^3\} = \{1, i, -1, -i\}$ of $\mathbb{C} - \{0\}$ is a cyclic group of order 4 (cf. 38.4, (iii) and (iv)).

It is helpful to note at the outset the following very simple point.

39.1 If G is a group such that (1) $x \in G$ and (2) every element of G is expressible as a power of x, then $G = \langle x \rangle$ (and thus G is cyclic).

Proof. Suppose that the conditions (1) and (2) hold for the group G. Because of (1), $\langle x \rangle$ is a subset of the group G. And, because of (2), every element of G belongs to the subset $\langle x \rangle$. Therefore $\langle x \rangle = G$.

If G is a cyclic group, an element $x \in G$ such that $\langle x \rangle$ is the whole of G is called a **generator** of G. (This fits in with the description of $\langle a \rangle$, where a is a group element, as the subgroup generated by a.)

By the meaning of the word "cyclic", for any cyclic group G there must be at least one generator of G; there may be more than one such generator, though it should not be imagined that every element of G must be a generator of G. For example, consider the case $G = \{2^n : n \in \mathbb{Z}\}$ $(=\langle 2 \rangle)$. Obviously 2 is a generator. So is $\frac{1}{2}$, since $\langle \frac{1}{2} \rangle$ is the whole of G. [This is clear from 39.1, since (1) $\frac{1}{2} \in G$ and (2) the arbitrary element 2^n of G is expressible as a power of $\frac{1}{2}$, viz. $(\frac{1}{2})^{-n}$.] On the other hand, the element $4 (=2^2)$ of G is not a generator of G, because the subgroup $\langle 4 \rangle$ is not the whole of G. [For example, $2 \notin \langle 4 \rangle$, since 2 cannot be expressed as an integral power of 4.]

If G is a finite cyclic group of order k and x is a generator of G, so that $G = \langle x \rangle$, then (making use of 38.4(iv)), we have

$$\text{period of } x = |\langle x \rangle| = |G| = k$$

Thus:

39.2 Every generator of a finite cyclic group of order k has period k.

Properties of cyclic groups
39.3 Every cyclic group is abelian.

Proof. Let G be a cyclic group, so that $G = \langle a \rangle$ for some element $a \in G$.

Let x, y be arbitrary elements of G. Since $G = \langle a \rangle$, $x = a^m$ and $y = a^n$ for some $m, n \in \mathbb{Z}$. Hence

$$xy = a^m . a^n = a^{m+n} = a^{n+m} = a^n . a^m = yx$$

Since x, y were arbitrary, this proves that G is abelian.

39.4 Every subgroup of a cyclic group is also cyclic.

Proof. Let $G = \langle a \rangle$ be an arbitrary cyclic group, and let H be a subgroup of G.

If H is the trivial subgroup (i.e. $\{1\}$), then $H = \langle 1 \rangle$, which is cyclic.

So we suppose henceforth that H is non-trivial. Consequently H contains a nonzero power of a, and (since $x \in H \Rightarrow x^{-1} \in H$) it follows that H contains at least one positive power of a. By the well-ordering principle (10.1), we may legitimately consider the lowest positive power of a that belongs to H: let this be the dth power of a. [We shall show that $H = \langle a^d \rangle$.]

Let x be an arbitrary element of H. Since $H \subseteq G = \langle a \rangle$, $x = a^n$ for some $n \in \mathbb{Z}$. By the division algorithm (11.1), we can write $n = qd + r$, for some $q, r \in \mathbb{Z}$ with $0 \leqslant r \leqslant d-1$; and, since a^n $(=x)$ and a^d belong to the subgroup H, we deduce that

$$a^n(a^d)^{-q} \in H, \quad \text{i.e. } a^{n-qd} \in H, \quad \text{i.e. } a^r \in H.$$

But $0 \leqslant r \leqslant d-1$, and a^d is the lowest positive power of a that belongs to H. Hence $r = 0$, and so $x = a^n = (a^d)^q$.

Thus every element of H is expressible as a power of a^d, which belongs to H. It follows, by 39.1, that $H = \langle a^d \rangle$.

The stated result now follows.

Worked problem. Let G be a finite cyclic group of order n, and let d be a divisor of n. Show that G has exactly one subgroup of order d.

Solution. Let x be a generator of G, so that, by 39.2, x has period n. Clearly $x^{n/d}$ has period d, and so G has a subgroup of order d, viz. the subgroup $H = \langle x^{n/d} \rangle$.

Let K be an arbitrary subgroup of G of order d. [We aim to prove that $K = H$.] By 39.4, K is cyclic: say $K = \langle x^t \rangle$ ($t \in \mathbb{Z}$). [Of course, since $K \subseteq G$, any generator of K belongs to G and is therefore a power of x.] By 39.2, x^t has period d and therefore $(x^t)^d = 1$, i.e. $x^{td} = 1$. Hence, by 38.4(ii), since x has period n, $n|dt$, i.e. $dt = qn$ for some $q \in \mathbb{Z}$. So $t = (n/d) \times q$, and hence $x^t = (x^{n/d})^q$. It follows that x^t belongs to $\langle x^{n/d} \rangle$, i.e. $x^t \in H$. Therefore, since H is a group, $\langle x^t \rangle \subseteq H$, i.e. $K \subseteq H$. As $|K| = |H|$, it follows that $K = H$.

Hence H is the one and only subgroup of order d in G.

EXERCISES ON CHAPTER SIX

N.B. The symbol G, wherever it occurs in the following questions, should be taken to denote a multiplicative group (arbitrary apart from restrictions imposed in the question).

1. Let G_1, G_2 be groups, and let multiplication be defined on $G_1 \times G_2$ by

$$(x_1, y_1)(x_2, y_2) = (x_1 x_2, y_1 y_2) \qquad (x_1, x_2 \in G_1, y_1, y_2 \in G_2).$$

Prove that (with respect to the multiplication just defined) $G_1 \times G_2$ is a group. (This group, to which we refer in some future exercises, is called the *direct product* of G_1, G_2.) Prove that the direct product $G_1 \times G_2$ is abelian iff both G_1 and G_2 are abelian.

2. Prove that, $\forall x, y \in G$, $\lambda_x \circ \lambda_y = \lambda_{xy}$, $\rho_x \circ \rho_y = \rho_{yx}$, $\lambda_x \circ \rho_y = \rho_y \circ \lambda_x$. (See §35 for the meanings of λ_a, ρ_a ($a \in G$).)

3. Let S be a finite multiplicative semigroup in which the cancellation laws hold (i.e. for $a, x, y \in S$, $ax = ay \Rightarrow x = y$ and $xa = ya \Rightarrow x = y$). Prove that S is a group.

Give an example of an infinite semigroup T such that the cancellation laws hold in T but T is not a group.

4. Suppose that $x, y \in G$ and that $(xy)^{-2} = y^{-2}x^{-2}$. Show that x and y commute.

5. In $GL(2, \mathbb{R})$, let H be the subset comprising all the matrices of the form $\begin{bmatrix} x & y \\ 0 & z \end{bmatrix}$

$(x, y, z \in \mathbb{R}, xz \neq 0)$, and let K be the subset comprising all the matrices $\begin{bmatrix} 1 & t \\ 0 & 1 \end{bmatrix} (t \in \mathbb{R})$.

Prove that H is a subgroup of $GL(2, \mathbb{R})$ and that K is a subgroup of H.

6. Let G be a group of permutations of the non-empty set Ω and, for each $t \in \Omega$, let $G_t = \{f \in G : f(t) = t\}$. Prove that, for each $t \in \Omega$, G_t is a subgroup of G. (G_t is called the *stabilizer* of t in G.)

7. (i) For $x, y \in G$, prove that x, y commute iff $xyx^{-1} = y$.
(ii) Prove that, for $n \geqslant 3$, $Z(S_n)$ is trivial.

8. Let x, $y \in G$. Prove that if $C(x) \subseteq C(y)$, then $xy = yx$. Prove further that $C(x) \subseteq C(y) \Leftrightarrow y \in Z(C(x))$.

9. Let H, K be subgroups of G such that neither $H \subseteq K$ nor $K \subseteq H$. Prove that $H \cup K$ is not a subgroup of G.

10. Give an example where H is a non-empty subset of G and H is closed under multiplication, but H is not a subgroup of G.

11. In $GL(2, \mathbb{R})$, let A be the matrix $\begin{bmatrix} 1 & -1 \\ 1 & 0 \end{bmatrix}$. By calculating the first few powers of A, determine the period of A.

12. Prove that, for $x \in G$, $x^2 = 1 \Leftrightarrow x = x^{-1}$.
Hence or otherwise prove that if every element of $G - \{1\}$ has period 2, then G is abelian.

13. Prove that $\forall x \in G$, x and x^{-1} have the same period. Deduce that if G is finite and has even order, then G contains an odd number of elements of period 2 (and, in particular, G contains at least one element of period 2).

14. If $x \in G - \{1\}$ and $x^{12} = x^2$, what periods are possible for x?

15. Let a, $x \in G$. Prove that, $\forall n \in \mathbb{N}$, $(xax^{-1})^n = xa^nx^{-1}$. Deduce that xax^{-1} has the same period as a.

Deduce further that (i) $\forall b$, $c \in G$, bc and cb have the same period, (ii) G cannot contain exactly two elements of period 2.

16. If x is a group element of period 20, what are the periods of x^4, x^{10}, x^{12}?

17. Draw up the multiplication table of the quaternion group. Show that it contains 6 elements of period 4 and just 1 element of period 2.

18. Let x be an element with odd period in G. Prove that $\exists y \in G$ s.t. $y^2 = x$.

19. Prove that the group formed by the nonzero congruence classes modulo 7 is cyclic.

20. Prove that the multiplicative group of positive rational numbers is non-cyclic.

21. Let $G_1 = \langle x \rangle$, $G_2 = \langle y \rangle$ be finite cyclic groups of orders m, n, respectively. Prove that, in $G_1 \times G_2$ (see question 1) the period of (x, y) is the l.c.m. of m and n (i.e. the least positive integer which both m and n divide). Deduce that the direct product of two cyclic groups of relatively prime orders is also cyclic.

22. Let $G = \langle x \rangle$, where x has period n, and let r denote a positive integer. Prove that x^r generates G iff hcf $(r, n) = 1$.

How many different generators has a cyclic group of order $2^k (k \in \mathbb{N})$?

How many different generators has an infinite cyclic group?

23. The non-trivial group G has no non-trivial proper subgroup. Prove that G is finite cyclic of prime order.

CHAPTER SEVEN

COSETS AND LAGRANGE'S THEOREM
ON FINITE GROUPS

40. Introduction

The highlight of this chapter is the theorem of Lagrange which states that the order of a subgroup of a finite group G is a factor of $|G|$. This simple fact is the foundation stone of the theory of finite groups. After proving the theorem in §43, we shall develop some straightforward applications of it in §44, e.g. we prove that every group of prime order is cyclic and that every group of order less than 6 is abelian. The ease with which we can prove these results in §44 is a measure of how significant an addition to our understanding of finite groups Lagrange's theorem is.

The proof of Lagrange's theorem itself comes easily enough, once we show (see diagram) that, given a finite group G with subgroup H, we can partition G into a number of equal-sized subsets, one of which is H. We arrive at the

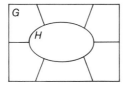

relevant partition of G by considering one or other of two equivalence relations called "left congruence modulo H" and "right congruence modulo H". These equivalence relations are defined and discussed in §42. The equivalence classes they determine turn out to be what we term *left* and *right* *cosets* of H. Cosets are introduced early in the chapter—in §41, where we prove the fact (needed for the proof of Lagrange's theorem) that each is the same size as the subgroup H.

D

41. Multiplication of subsets of a group

Throughout this section, G will denote a group.

We begin by giving a meaning to the product XY, where X, Y are subsets of G.

Definition. Let X, Y be subsets of the group G. Then XY means the subset

$$\{xy : x \in X, y \in Y\}$$

of G, i.e. the set of all elements of G that can be expressed as

(an element of X). (an element of Y)   [in that order].

XY is $\emptyset$ if X or Y is $\emptyset$.

Illustration. In the group of symmetries of the equilateral triangle (whose multiplication table was given in §34(5)), let $X = \{1, p\}$ and $Y = \{a, c\}$. Then

$$XY = \{1a, 1c, pa, pc\}$$

Now $1a, 1c, pa, pc$ are respectively equal to a, c, c, b. Hence in this case XY is the 3-element subset $\{a, b, c\}$.

It is worth while noting that:

41.1 With respect to multiplication of subsets of G, $\mathscr{P}(G)$ is a semigroup.

Proof. It is clear that $\mathscr{P}(G)$ is non-empty and is closed under multiplication (of subsets of G).

Let X, Y, Z be three arbitrary subsets of G. For $g \in G$, we have:

$$
\begin{aligned}
g \in (XY)Z &\Leftrightarrow g = tz \text{ for some } t \in XY, z \in Z \\
&\Leftrightarrow g = (xy)z \text{ for some } x \in X, y \in Y, z \in Z \\
&\Leftrightarrow g = x(yz) \text{ for some } x \in X, y \in Y, z \in Z \quad \text{(by the associativity} \\
&\qquad\qquad\qquad\qquad\qquad\qquad\qquad\qquad\qquad\quad \text{of the multiplication defined on } G) \\
&\Leftrightarrow g = xu \text{ for some } x \in X, u \in YZ \\
&\Leftrightarrow g \in X(YZ).
\end{aligned}
$$

Hence $X(YZ) = (XY)Z$.

This proves that the multiplication defined on $\mathscr{P}(G)$ is associative; and it follows that $(\mathscr{P}(G), .)$ is a semigroup.

As an important special case, let us consider the product XY (where X, $Y \subseteq G$) when either X or Y is a singleton (i.e. a one-membered set).

If $X = \{a\}$ (a being an element of G), the product of the subsets is $\{a\}Y$, which we normally write simply as aY. Thus

$$
\begin{aligned}
aY &= \{ay : y \in Y\} \\
&= \text{set of elements of } G \text{ obtainable from elements of } Y \text{ by left multiplication by } a \\
&= \lambda_a(Y).
\end{aligned}
$$

[See §35 for the meaning of λ_a.]

Similarly, if $Y = \{b\}$ (b being an element of G), the product XY is $X\{b\}$, which we normally write simply as Xb. Thus

$Xb = \{xb : x \in X\}$
    = set of elements of G obtainable from elements of X by right multiplication by b
    $= \rho_b(X)$.

Since (see 35.8) both λ_a and ρ_b are injective (for all $a, b \in G$), we deduce by 22.2 the following result about the sizes of the subsets aY, Xb:

41.2 (i) If Y is a finite subset of G and $a \in G$, then $|aY| = |Y|$.
    (ii) If X is a finite subset of G and $b \in G$, then $|Xb| = |X|$.

Cosets
We are going to be much interested in the subsets xH and Hx, where x is an element of the group G and H is a subgroup of G. Such subsets are termed *cosets* of H. More precisely:

Definition. If H is a subgroup of the group G, then (1) for each $x \in G$, the subset xH of G is called the **left coset** of H determined by x, and (2) for each $x \in G$, the subset Hx of G is called the **right coset** of H determined by x.

As a particular case of 41.2, note that:

41.3 If H is a finite subgroup of the group G, then each left or right coset of H has the same order as H.

42. Another approach to cosets

Throughout this section, G will denote a group and H will denote a subgroup of G.

We are about to define two equivalence relations on G. It will turn out that the equivalence classes determined by these relations are in one case the left cosets of H and in the other case the right cosets of H.

Definitions. (1) A relation called **left congruence modulo H** is defined on G by

$$x \equiv^l y \,(\mathrm{mod}\ H) \Leftrightarrow x^{-1}y \in H \qquad (x, y \in G)$$

["$x \equiv^l y \,(\mathrm{mod}\ H)$" is read "$x$ is left congruent to y modulo H".]

    (2) A relation called **right congruence modulo H** is defined on G by

$$x \equiv^r y \,(\mathrm{mod}\ H) \Leftrightarrow xy^{-1} \in H \qquad (x, y \in G)$$

["$x \equiv^r y \,(\mathrm{mod}\ H)$" is read "$x$ is right congruent to y modulo H".]

42.1 Left and right congruence modulo H are equivalence relations on G.

Proof. We give the proof for left congruence, the proof for right congruence being similar.

(1) For every $x \in G$, $x \equiv^l x \,(\text{mod } H)$ because $x^{-1}x = 1$ and $1 \in H$ (since H is a subgroup). Hence left congruence (modulo H) is reflexive.

(2) For $x, y \in G$,

$$
\begin{aligned}
x \equiv^l y \,(\text{mod } H) &\Rightarrow x^{-1}y \in H \\
&\Rightarrow (x^{-1}y)^{-1} \in H \qquad \text{(since } H \text{ is a subgroup)} \\
&\Rightarrow y^{-1}x \in H \qquad \text{(by reversal rule, 35.6)} \\
&\Rightarrow y \equiv^l x \,(\text{mod } H).
\end{aligned}
$$

Hence left congruence (modulo H) is symmetric.

(3) For $x, y, z \in G$,

$$
\begin{aligned}
x \equiv^l y \,(\text{mod } H) \text{ and } y \equiv^l z \,(\text{mod } H) &\Rightarrow x^{-1}y \in H \text{ and } y^{-1}z \in H \\
&\Rightarrow (x^{-1}y)(y^{-1}z) \in H \\
&\qquad \text{(since } H \text{ is a subgroup)} \\
&\Rightarrow x^{-1}z \in H \\
&\Rightarrow x \equiv^l z \,(\text{mod } H).
\end{aligned}
$$

Hence left congruence (modulo H) is transitive.

It follows that left congruence (modulo H) is an equivalence relation.

With reference to the definitions of left and right congruence (modulo H) note that $x^{-1}y$, xy^{-1} are, so to speak, measures of how x, y differ multiplicatively, i.e. they are the answers to the questions

$$
y = x \,.\, (\text{WHAT?}), \qquad x = (\text{WHAT?}) \,.\, y
$$

respectively. So, when we assert that elements x, y are left or right congruent to each other modulo H, we are saying that they differ from each other multiplicatively by an element of H (on the right or left, respectively). This can be indicated more formally by giving various obviously equivalent versions of the meanings of $x \equiv^l y \,(\text{mod } H)$ and $x \equiv^r y \,(\text{mod } H)$, as follows.

42.2 For $x, y \in G$,

(i) $x \equiv^l y \,(\text{mod } H) \Leftrightarrow x^{-1}y \in H$
$\qquad\qquad\qquad \Leftrightarrow x^{-1}y = h \text{ for some } h \in H$
$\qquad\qquad\qquad \Leftrightarrow y = xh \text{ for some } h \in H;$

(ii) $x \equiv^r y \,(\text{mod } H) \Leftrightarrow xy^{-1} \in H$
$\qquad\qquad\qquad \Leftrightarrow xy^{-1} = h \text{ for some } h \in H$
$\qquad\qquad\qquad \Leftrightarrow x = hy \text{ for some } h \in H$

In an abelian group, the distinction between left and right congruence modulo H disappears, and we can refer to both relations simply as "congruence modulo H".

In an additive abelian group G, the condition for elements x, y to be congruent to each other modulo the subgroup H is

$$x - y \in H$$

i.e. x, y differ (additively) by an element of H.

At this point, the student is likely to realize that he has already met a particular case of congruence modulo a subgroup. In discussing integers, when we say "$x \equiv y \pmod{m}$" (x, y, m being integers with m positive), we are stating that x and y are congruent to each other modulo the subgroup M of $(\mathbb{Z}, +)$ comprising all the integral multiples of m (i.e. that x, y differ additively by an element of that subgroup M).

Returning to the general case of a multiplicative group G with subgroup H, we come to the identification of the equivalence classes determined by left and right congruence modulo H.

42.3 For each $x \in G$,

(i) the left congruence class of $x \pmod{H}$ is the left coset xH;

(ii) the right congruence class of $x \pmod{H}$ is the right coset Hx.

Proof. Let x be an arbitrary element of G, and let C_x be the left congruence class of $x \pmod{H}$. Then, for $t \in G$,

$$t \in C_x \Leftrightarrow x \equiv^l t \pmod{H} \qquad \text{(by the definition of "equivalence class")}$$
$$\Leftrightarrow t = xh \text{ for some } h \in H \qquad \text{(see 42.2)}$$
$$\Leftrightarrow t \in xH \qquad \text{(by meaning of } xH\text{)}.$$

Hence $C_x = xH$.

This proves part (i). The proof of part (ii) is similar.

It should be noted that 42.3 tells us both that each congruence class is a coset and that each coset is a congruence class! Knowing this, we can deduce important results about cosets from general theorems about equivalence relations, e.g. from 17.5 we deduce:

42.4 The set of left cosets of H is a partition of G; and so is the set of right cosets of H.

And from 17.6 it follows that:

42.5 For each $x \in G$,

(i) the one and only left coset (of H) to which x belongs is xH;

(ii) the one and only right coset (of H) to which x belongs is Hx.

The remark following 17.3 and the clarificatory note following 17.5 are also highly relevant to cosets. In particular, it must be borne in mind that each coset of H has more than one label of the form xH (or Hx) provided it contains more than one element; e.g. if a left coset of H contains the elements $x_1, x_2, x_3, \ldots$, then that left coset is equal to $x_1 H$ and to $x_2 H$ and to $x_3 H \ldots$. Correspondingly, due care must be exercised in applying 42.5, e.g. if we are given that $x \in yH$ (where $x, y \in G$), then 42.5 entitles us to say that $yH = xH$, but it does not entitle us to infer that $x = y$.

The diagram below illustrates the partitions described in 42.4. It also indicates certain simple facts about cosets, e.g. (a) in each partition H itself is one of the cosets (since $1H = H$ and $H1 = H$), and (b) every coset has the same size (see 41.3).

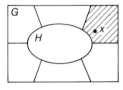

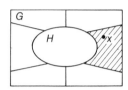

(i) partition of G into left cosets: shaded area represents xH.

(ii) partition of G into right cosets: shaded area represents Hx.

A re-reading of §40 will reveal that the road to proof of Lagrange's theorem is now open. Before taking that road, however, let us pause to complete the job of recording consequences of 42.3. The point to be made is that 42.3 enables us to translate a condition involving cosets of H into an equivalent condition involving elements of G (but not mentioning cosets), and vice versa. Consider, for instance, the condition $x \in yH$ $(x, y \in G)$. Because yH is the left congruence class of y (mod H), the condition $x \in yH$ is equivalent to $x \equiv^l y$ (mod H), i.e. to $x^{-1}y \in H$. By similar thinking, the student should be able to prove the following list of equivalences.

42.6 For $x, y \in G$,

(1L) $x \in yH \Leftrightarrow x^{-1}y \in H$ (1R) $x \in Hy \Leftrightarrow xy^{-1} \in H$

(2L) $xH = yH \Leftrightarrow x^{-1}y \in H$ (2R) $Hx = Hy \Leftrightarrow xy^{-1} \in H$

(3L) $xH = H \Leftrightarrow x \in H$ (3R) $Hx = H \Leftrightarrow x \in H$

(4L) x, y belong to the same left coset of $H \Leftrightarrow x^{-1}y \in H$

(4R) x, y belong to the same right coset of $H \Leftrightarrow xy^{-1} \in H$

As the worked examples below indicate, it is important to have all the information in 42.6 at one's fingertips. Some students may find it helpful to note that (2L), (2R) can be reduced to the easily remembered (3L), (3R) through the observation that (for $x, y \in G$)

$$xH = yH \Leftrightarrow x^{-1}(xH) = x^{-1}(yH) \Leftrightarrow H = (x^{-1}y)H$$

and

$$Hx = Hy \Leftrightarrow (Hx)y^{-1} = (Hy)y^{-1} \Leftrightarrow H(xy^{-1}) = H$$

Worked examples

1. Take G to be the group $\{1, p, q, a, b, c\}$ of symmetries of the equilateral triangle (as in §34(5)), and take H to be the subgroup $\langle a \rangle = \{1, a\}$. Determine explicitly how G partitions into left and right cosets of H in this case.

Solution. The left cosets of H are as follows:

$$H \text{ itself} = \{1, a\} = 1H = aH$$
$$bH = \{b1, ba\} = \{b, q\} = qH$$
$$cH = \{c1, ca\} = \{c, p\} = pH$$

[The left-hand diagram below illustrates the partition of G into these left cosets. In the above analysis, notice how, as soon as we know that a subset X is a left coset and that it contains the element t, we can at once conclude (by 42.5) that $X = tH$. E.g. having found above that $bH = \{b, q\}$, we know that $\{b, q\}$ is a left coset containing q and can infer straight away that $\{b, q\}$ equals qH.]

'The right cosets of H are as follows:

$$H \text{ itself} = \{1, a\} = H1 = Ha$$
$$Hb = \{1b, ab\} = \{b, p\} = Hp$$
$$Hc = \{1c, ac\} = \{c, q\} = Hq$$

The diagram on the right illustrates the partition of G into these right cosets.

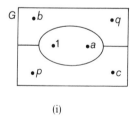 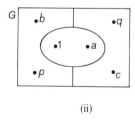

(i)   (ii)

[*Note*: as this case illustrates, the partitions of G into left and right cosets may be different.]

2. Let a be a fixed element of the group G and let $C(a)$ be the centralizer of a in G (see §37(1)). Prove that, for $g, h \in G$, $gag^{-1} = hah^{-1} \Leftrightarrow g, h$ belong to the same left coset of $C(a)$.

Solution. For $g, h \in G$,

$$gag^{-1} = hah^{-1} \Leftrightarrow g^{-1}(gag^{-1})h = g^{-1}(hah^{-1})h \qquad \text{(right to left impli-}$$
$$\text{cation by cancellation laws)}$$
$$\Leftrightarrow ag^{-1}h = g^{-1}ha$$
$$\Leftrightarrow g^{-1}h \in C(a) \qquad \text{(by definition of } C(a))$$
$$\Leftrightarrow g, h \text{ belong to same left coset of } C(a) \qquad \text{(see 42.6).}$$

3. Let $G = \langle a \rangle$ be an infinite cyclic group, and let H be the subgroup $\langle a^s \rangle$, where s is a given positive integer. Show that

$$H, aH, a^2H, \ldots, a^{s-1}H \qquad \qquad (*)$$

is a complete repetition-free list of the cosets of H in G.

Solution. The problem subdivides into proving (i) that the list (*) is complete, i.e. that every element of G lies in one of the cosets listed, and (ii) that the list (*) is repetition-free.

(i) Let x be an arbitrary element of G. Then $x = a^n$ for some $n \in \mathbb{Z}$. By the division algorithm, $n = qs + r$ for some $q, r \in \mathbb{Z}$ with $0 \leqslant r \leqslant s-1$. Hence we have

$$x = a^{qs+r} = a^r(a^s)^q = a^r \cdot (\text{an element of } H)$$

and thus $x \in a^r H$, which (since $0 \leqslant r \leqslant s-1$) is one of the cosets in the list (*).

It follows that every element of G belongs to (at least) one of the cosets in the list (*).

(ii) Suppose (with a view to obtaining a contradiction) that

$$a^i H = a^j H \text{ with } 0 \leqslant i < j \leqslant s-1$$

Then (see 42.6) $(a^i)^{-1}a^j \in H$, i.e. $a^{j-i} \in H$, and so $a^{j-i} = a^{ks}$ for some $k \in \mathbb{Z}$. Since a has infinite period, we deduce (by 38.2) that $j - i = ks$—a contradiction since $0 < j - i < s$.

From this contradiction it follows that the list (*) is repetition-free.

The desired result is now established.

43. Lagrange's theorem

One further definition is necessary before we come to the full detail of Lagrange's theorem.

Definition. If H is a subgroup of the group G, the (left) **index** of H in G is the number of different left cosets of H in G.

Whenever this number is finite, we denote it by $|G : H|$.

43.1 (Lagrange's theorem) Let G be a finite group and let H be a subgroup of

G. Then $|G| = |G:H| \times |H|$. In particular, therefore, both $|H|$ and $|G:H|$ are factors of $|G|$.

Proof. Let $|H| = m$ and $|G:H| = k$. Consider the partition of G into left cosets of H (see diagram).

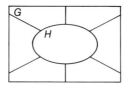

Since $|G:H| = k$, there are in the partition k different cosets and, by 41.3, each contains m elements. Hence altogether G contains km elements,

i.e. $$|G| = km = |G:H| \times |H|$$

which is the result.

Worked examples

1. In a group G, H and K are different subgroups, both of order p, where p is prime. Show that $H \cap K = \{1\}$.

Solution. $H \cap K$ is a subgroup of G (by 36.6) and indeed, since $H \cap K \subseteq H$, $H \cap K$ is a subgroup of H. So, by Lagrange's theorem, $|H \cap K| \mid |H|$, i.e. $|H \cap K| \mid p$. Therefore, since p is prime, $|H \cap K| = 1$ or p.

Since H, K are unequal sets, each of order p, it is clear that H is not contained in K and therefore (cf. 7.3(ii)) $H \cap K \neq H$. Hence $H \cap K \subset H$ and so $|H \cap K| < |H| = p$.

It follows that $|H \cap K| = 1$. Hence, since $H \cap K$, being a subgroup, must contain the element 1, we deduce that $H \cap K = \{1\}$.

2. Let G be a finite group, and let H, K be subgroups of G such that $K \subseteq H$. Show that $|G:H|$ is a factor of $|G:K|$.

Solution. By Lagrange's theorem, $|G:H| = |G|/|H|$ and $|G:K| = |G|/|K|$. Hence

$$|G:K| = (|H|/|K|) \times |G:H|$$

But, since K is a subgroup of H, it follows from Lagrange's theorem that $|H|/|K|$ is an integer. So the last equation gives the result.

The rest of this section is devoted to a point arising from Lagrange's theorem and the way in which we proved it. Consider a group G with subgroup

H. In 43.1 we showed, by considering the partition of G into left cosets of H, that, if G is finite,

number of different left cosets of H in $G = |G|/|H|$.

Clearly, if instead we had considered the partition of G into right cosets of H, the same argument would have led us to the conclusion that, if G is finite,

number of different right cosets of H in $G = |G|/|H|$.

Therefore, if G is finite, the numbers of different left and right cosets of the subgroup H in G are equal. We now use a different method to prove that this fact remains true whenever there are finitely many different left cosets of H in G, even if G is infinite.

43.2 Suppose that H is a subgroup of the group G and that there are exactly r different left cosets of H in G; and let $a_1 H, a_2 H, \ldots, a_r H$ be a complete repetition-free list of them $(a_1, a_2, \ldots, a_r \in G)$. Then there are exactly r different right cosets of H in G, and $Ha_1^{-1}, Ha_2^{-1}, \ldots, Ha_r^{-1}$ is a complete repetition-free list of them.

Proof. By hypothesis, each element of G belongs to precisely one of the left cosets $a_1 H, a_2 H, \ldots, a_r H$.

Let x be an arbitrary element of G. We note that, for $1 \leqslant i \leqslant r$,

$$x \in Ha_i^{-1} \Leftrightarrow x(a_i^{-1})^{-1} \in H \qquad \text{(see 42.6)}$$
$$\Leftrightarrow xa_i \in H$$
$$\Leftrightarrow (x^{-1})^{-1}a_i \in H$$
$$\Leftrightarrow x^{-1} \in a_i H \qquad \text{(42.6 again)}.$$

Since (by the opening sentence of the proof) the condition $x^{-1} \in a_i H$ is true for precisely one value of i, it follows that the equivalent condition $x \in Ha_i^{-1}$ is true for precisely that same one value of i.

This proves that each element of G belongs to precisely one of the right cosets $Ha_1^{-1}, Ha_2^{-1}, \ldots, Ha_r^{-1}$, and the stated result follows.

44. Some consequences of Lagrange's theorem

The consequences we discuss in this section have largely t do with periods of elements and stem from the following corollary of Lagrange's theorem.

44.1 In a finite group G, the period of each element divides $|G|$.

Proof. Let x be an arbitrary element of the finite group G. By 38.4(iv), the period of x equals the order of the subgroup $\langle x \rangle$ and is, therefore, by 43.1, a factor of $|G|$. This proves the result.

We use 44.1 to prove the next result, which is a simple example of something of great interest in finite group theory—namely the extent to which the properties of a finite group are determined by the prime decomposition of its order. (Another example, which we shall establish in chapter 8, is that every group whose order is the square of a prime is abelian.)

44.2 Every group of prime order is cyclic. Moreover, in a group of prime order p, every non-identity element (i.e. every element other than 1) has period p.

Proof. Let G be a group of order p, where p is prime.

Consider an arbitrary element x of the non-empty set $G - \{1\}$. By 44.1, the period of x is a factor of the prime p; but (cf. 38.1) the period of x is not 1, since $x \neq 1$. Therefore the period of x is p. Hence the subgroup $\langle x \rangle$ has order p and is therefore the whole of G.

Thus G is cyclic, every non-identity element of it having period p.

We are now in a position to prove:

44.3 Every group of order less than 6 is abelian.

For this purpose, we re-state a result set as an exercise on chapter 6:

44.4 A group in which every non-identity element has period 2 is abelian.

Proof of 44.3. A group of order 1 (i.e. a trivial group) is clearly abelian. And, since 2, 3, 5 are primes, it follows from 44.2 that every group of order 2, 3 or 5 is cyclic and hence (by 39.3) abelian. So it remains to prove that every group of order 4 is abelian.

Let then G be a group of order 4; and suppose, with a view to obtaining a contradiction, that G is non-abelian. No element of G can have period 4: for if x $(\in G)$ had period 4, $\langle x \rangle$ would be the whole of G and so G would be cyclic and therefore abelian. But, by 44.1, every element of G has period dividing 4, i.e. has period 1, 2 or 4. Hence every non-identity element of G has period 2, and so, by 44.4, G is abelian—a contradiction.

From this contradiction the stated result follows.

Four-groups

As the proof of 44.3 might lead one to suspect, there do exist groups of order 4 in which every non-identity element has period 2. Any group of order 4 with this property is called a **Klein four-group** (or simply a **four-group**).

There is an important sense in which all Klein four-groups deserve to be described as "the same" or "the same in structure". In particular, apart from variations in labelling the elements, there is only one possible multiplication table for a four-group.

44.5 For a four-group whose non-identity elements are denoted by a, b, c, the multiplication table can only be:

	1	a	b	c
1	1	a	b	c
a	a	1	c	b
b	b	c	1	a
c	c	b	a	1

Outline of proof. In constructing the multiplication table, one can begin by filling in the row and column headed 1 and then entering the information $a^2 = b^2 = c^2 = 1$ (this being the defining feature of a four-group). One then finds that, by virtue of 35.9, there is only one way to complete the table.

So far no justification of the assertion that four-groups exist has been given. That omission is fully repaired by the following two illustrations.

Illustration 1. Let G consist of the following symmetries of the rectangle drawn below:

$$1 = \text{identity mapping from the plane to itself,}$$

$$a = \text{reflection in } x\text{-axis}, \qquad b = \text{reflection in } y\text{-axis,}$$

$$c = \text{half-turn (i.e. rotation through } 180°) \text{ about } O.$$

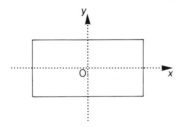

It is easily verified that:

$$a^2 = b^2 = c^2 = 1; \quad ab = ba = c; \quad bc = cb = a; \quad ca = ac = b$$

Hence it is clear that G, which is a finite non-empty subset of the group of all isometries of the plane, is closed under multiplication. Therefore, by 36.5, G is a group. And, since $a^2 = b^2 = c^2 = 1$, G is a four-group.

Illustration 2. Let G_1 consist of the following elements of S_4:

$$1_1 = \text{identity permutation of } \{1, 2, 3, 4\},$$

$$a_1 = \begin{pmatrix} 1 & 2 & 3 & 4 \\ 2 & 1 & 4 & 3 \end{pmatrix}, \quad b_1 = \begin{pmatrix} 1 & 2 & 3 & 4 \\ 3 & 4 & 1 & 2 \end{pmatrix}, \quad c_1 = \begin{pmatrix} 1 & 2 & 3 & 4 \\ 4 & 3 & 2 & 1 \end{pmatrix}$$

Then (imitating the method employed in illustration 1) one can show that

G_1 is a four-group. Its multiplication table is exactly the same as the table in 44.5, except that all the elements have subscript 1s in their labels.

Worked examples

1. Let G be a non-abelian group of order 10. Prove that G contains at least one element of period 5.

Solution. By 44.1, the period of each element of G is a factor of 10, i.e. is 1 or 2 or 5 or 10. However, no element of G has period 10 (since G is non-abelian and therefore non-cyclic); and only the identity has period 1. Hence each non-identity element has period 2 or 5. But, if all the non-identity elements had period 2, G would be abelian (cf. 44.4)—which is not the case. Therefore, at least one element of G has period 5.

2. Let G be a group of order 27. Prove that G contains a subgroup of order 3.

Solution. Choose any element $x \in G - \{1\}$. Since $x \neq 1$ and since the period of x divides $|G|$ $(=27)$, it follows that the period of x is 3 or 9 or 27. Correspondingly, x or x^3 or x^9 has period 3, and $\langle x \rangle$ or $\langle x^3 \rangle$ or $\langle x^9 \rangle$ is a subgroup of G of order 3.

Remark. It must not be supposed that there is a general converse of Lagrange's theorem that says that if d is a factor of the order of a finite group G, then G must have a subgroup of order d. For example, the group S_5 has order 5!, i.e. 120 which has 15 as a factor, but there is no subgroup of order 15 in S_5—a fact whose proof is one of the matters to be dealt with in one of the exercises following chapter 9. However, it is true (though proof is beyond the scope of this book) that if d is a prime power (i.e. an integer of the form p^r, where p is a prime and $r \in \mathbb{N}$) and d divides the order of a finite group G, then G has at least one subgroup of order d. This is part of the information provided by a set of beautiful and remarkable theorems proved in the 1870s by the Norwegian mathematician Sylow, of which students will hear a lot more if they pursue the study of finite group theory beyond the level of this textbook.

EXERCISES ON CHAPTER SEVEN

N.B. As in the exercises on chapter 6, the symbol G, wherever it occurs in the following questions, should be taken to denote a multiplicative group (arbitrary apart from restrictions imposed in the question).

1. Suppose that $a \in G$ and that $X = \{1, a^2\}$, $Y = \{1, a, a^3\}$. How many different elements has the subset XY (i) if a has period 4, (ii) if a has period 6?

2. Prove that if H is a subgroup of G, then $HH = H$.

3. Suppose that G is finite and that X is a subset of G such that $|X| > \frac{1}{2}|G|$. Prove that $XX = G$.

4. Let X, Y, Z be subsets of G. (i) Show that

$$X(Y \cap Z) \subseteq XY \cap XZ. \tag{*}$$

(ii) By considering the case $X = \{a, a^{-1}\}$, $Y = \{a\}$, $Z = \{a^{-1}\}$, where a is an element of G such that $a^2 \neq 1$, show that it is possible to have $X(Y \cap Z) \neq XY \cap XZ$. (iii) Show that equality holds in (*) when $|X| = 1$, i.e. that, for Y, $Z \subseteq G$ and $x \in G$, $x(Y \cap Z) = xY \cap xZ$.

5. Let H be a subgroup of G, and let $g \in G$. Prove that gHg^{-1} is also a subgroup of G and that, if H is finite, $|gHg^{-1}| = |H|$.

6. For $X \subseteq G$, one defines X^{-1} to mean $\{x^{-1} : x \in X\}$. Show that (i) for all X, $Y \subseteq G$, $(XY)^{-1} = Y^{-1}X^{-1}$, (ii) if L is a subgroup of G, then $L^{-1} = L$, (iii) if X is a non-empty subset of G, then X is a subgroup iff $X^{-1}X \subseteq X$.

Deduce that, if H, K are subgroups of G, then HK is a subgroup iff $HK = KH$.

7. Translate each of the following into a statement not involving cosets. (Here H denotes a subgroup of G, and x, y, a, b, c, s, t denote elements of G.) (i) $x \in yH$; (ii) a, b belong to the same right coset of H; (iii) $Hc = H$; (iv) $sH = tH$.

8. Suppose that H, L are subgroups of G with $H \subseteq L$. Show that, for each $x \in G$, either $xH \subseteq L$ or $xH \cap L = \emptyset$.

9. Suppose that $aH \cap bK \neq \emptyset$, where H, K are subgroups of G and a, $b \in G$. Prove that $aH \cap bK$ is a left coset of the subgroup $H \cap K$. (*Hint*: see last part of question 4.)

10. Suppose that H and K are subgroups of G and that $xH \subseteq yK$ for certain elements x, $y \in G$. Prove that $H \subseteq K$.

11. Let H, K be subgroups of G. Show that if $a_1, a_2, \ldots, a_n \in H$ and if the cosets $a_1(H \cap K), a_2(H \cap K), \ldots, a_n(H \cap K)$ of $H \cap K$ in H are all different, then the cosets $a_1K, a_2K, \ldots, a_nK$ of K in G are all different. Deduce that if $|G:K|$ is finite, $|H:H \cap K| \leqslant |G:K|$.

Suppose now that $|G:H| = 8$, $|G:K| = 9$, and $|H \cap K| = 5$. Show that G is finite, and determine its order. (Note that $|G:H \cap K| = |G:H| \times |H:H \cap K|$—cf. worked example 2 in §43.)

12. Let H, K be finite subgroups of G. Prove that, for $h_1, h_2 \in H$, $h_1K = h_2K \Leftrightarrow h_1, h_2$ belong to the same left coset of $H \cap K$ in H. Hence or otherwise prove that (whether or not HK is a subgroup)

$$|HK| = |H||K|/|H \cap K|.$$

Deduce that, for all $g \in G$, $|HgK| = |H||K|/|H \cap gKg^{-1}|$.

13. In a finite group G there are subgroups H, K of orders 25, 36, respectively. Prove that (i) $H \cap K = \{1\}$, (ii) $900 \mid |G|$.

14. Suppose that H, K are unequal subgroups of G, each of order 16. Prove that $24 \leqslant |H \cup K| \leqslant 31$.

15. Suppose that G is finite and that H, K are subgroups of G such that $K \subset H$ and $|G:K|$ is prime. Prove that $H = G$.

16. Let G be finite. Prove that, for each $x \in G$, the period of x divides $|C(x)|$.

17. Let G be a group of permutations of the non-empty set Ω, let $t \in \Omega$, and let G_t be the stabilizer of t in G (see exercise 6 on chapter 6). Show that each left coset of G_t in G comprises all the elements of G that map t to some particular element of Ω. Deduce that, if G is finite, $|G:G_t|$ equals the number of different images that t has under elements of G, and conclude that this number divides $|G|$.

18. (i) Suppose that G is finite. Prove that, $\forall x \in G$, $x^{|G|} = 1$.

(ii) Let p be a prime number. Assuming that the nonzero congruence classes modulo p form a group with respect to multiplication, use part (i) to prove Fermat's theorem (see §19) that if $n \in \mathbb{Z}$ and $p \nmid n$, then $n^{p-1} \equiv 1 \pmod{p}$.

19. Suppose that the elements x, y of G have relatively prime (finite) periods m, n, respectively. Show that $\langle x \rangle \cap \langle y \rangle = \{1\}$. Deduce that, if x, y commute with each other, then the period of xy is mn.

20. Prove that a non-abelian group of order 8 contains at least one element of period 4.

21. Suppose that G has exactly m subgroups of order p (p being prime). Show that the total number of elements of period p in G is $m(p-1)$.

Deduce the following results. (i) A non-cyclic group of order 55 has at least one element of period 5 and at least one element of period 11. (ii) A non-cyclic group of order p^2 (p prime) has altogether $p+3$ subgroups.

22. Suppose that G is finite and abelian and that G contains more than 2 elements of period 3. By considering an appropriate subgroup, show that $9 \mid |G|$.

State a generalization with 3 replaced by an arbitrary prime p.

Deduce that if G is abelian of order pq, where p, q are distinct primes, then G is cyclic.

CHAPTER EIGHT

HOMOMORPHISMS, NORMAL SUBGROUPS, AND QUOTIENT GROUPS

45. Introduction

In the context of group theory, the word *homomorphism* means a mapping from one group to another which, so to speak, respects the operations of multiplication defined on the two groups. To be more explicit, a homomorphism from a group G to a group G_1 is a mapping $\theta : G \to G_1$ which satisfies

$$\theta(xy) = \theta(x)\theta(y) \quad \text{for all } x, y \in G$$

As students work through this chapter, they will come to see that it is such mappings that enable us to express and explore relationships between groups. For example, as will be seen in §46, the existence of a bijective homomorphism from a group G to a group G_1 tells us that the groups G, G_1 are the same in structure, the same in all respects that abstract algebra is concerned with (like the two illustrations of four-groups given in §44).

A discussion of homomorphisms in general (not just bijective ones) and of their basic properties is given in §47.

Sections 48 to 52 are devoted to different but related topics. In §49 we define what we mean by a *normal* subgroup of a given group G, and in §50 we show that if K is such a subgroup of G, we can make the set of left cosets of K in G into a group—a procedure foreshadowed in §18 when we made $\mathbb{Z}_m$ into an algebraic system. One of the first-fruits of this idea is a proof of the fact that a group whose order is the square of a prime must be abelian (§51).

46. Isomorphic groups

In geometry the recognition of congruent triangles is important because two congruent triangles, though in general distinct objects with different positions in space, have identical internal properties (sides of the same length, same area, etc.). Analogously, we saw in §44 that there is a sense in which all four-groups

104

(whether their elements are matrices or mappings or anything else) are the same: they have, we could say, the same internal properties, the same algebraic structure. This was most clearly seen in the result 44.5.

Mathematicians describe two algebraic systems with the same algebraic structure as "isomorphic to each other". By derivation *isomorphic* means *of equal form*. This idea is of great importance in the study of algebraic systems of all kinds. So far, our formulation of it has been rather vague, and our immediate aim must be to move to a precise definition of what we mean by *isomorphic groups*.

Let us begin by considering two finite groups G, G_1. In the case of a finite group, all the information about its "algebraic structure" is contained in the multiplication table. Accordingly, what we require is a concise criterion for G, G_1 to have (once elements have been re-labelled if necessary) the same multiplication table. When that is the case, there must be a bijection $\theta: G \rightarrow G_1$, associating each element of G with the element of G_1 that plays the same role in the multiplication table of G_1; and, because of the similarity of the multiplication tables, we must have

$$xy = z \text{ in } G \Rightarrow \theta(x)\theta(y) = \theta(z) \text{ in } G_1$$

so that

$$\forall x, y \in G, \quad \theta(xy) = \theta(x)\theta(y) \qquad (*)$$

It can also be seen that, conversely, if a bijection $\theta: G \rightarrow G_1$ satisfying (*) is given, then, by re-labelling $\theta(x)$ as x for each $x \in G$, we can arrive at the situation where G, G_1 have identical multiplication tables.

Since these considerations extend to infinite groups if we interpret "multiplication table" to mean the totality of information about multiplication in the group, it becomes apparent that it makes sense to give the following formal definition of *isomorphic groups*.

Definition. Let G, G_1 be groups. Then G is said to be **isomorphic** to G_1 iff there exists a bijection θ from G to G_1 such that

$$\forall x, y \in G, \quad \theta(xy) = \theta(x)\theta(y) \qquad (*)$$

Moreover, a mapping θ from G to G_1 that is bijective and satisfies (*) is called an **isomorphism** from G to G_1. The standard notation for "G is isomorphic to G_1" is

$$G \cong G_1$$

46.1 For any set S of groups, $\cong$ is an equivalence relation on S.

Sketch of proof. (1) For every $G \in S$, $G \cong G$ because i_G is an isomorphism from G to G.

(2) For $G_1, G_2 \in S$,

$G_1 \cong G_2 \Rightarrow$ there is an isomorphism $\theta : G_1 \to G_2$
$\quad\quad\quad\quad \Rightarrow$ there is an isomorphism (viz. θ^{-1}) from G_2 to G_1
$\quad\quad\quad\quad \Rightarrow G_2 \cong G_1$.

(3) For $G_1, G_2, G_3 \in S$,

$G_1 \cong G_2$ and $G_2 \cong G_3 \Rightarrow$ there are isomorphisms $\theta : G_1 \to G_2$, $\phi : G_2 \to G_3$
$\quad\quad\quad\quad\quad\quad\quad\quad\quad \Rightarrow$ there is an isomorphism (viz. $\phi \circ \theta$) from G_1 to G_3
$\quad\quad\quad\quad\quad\quad\quad\quad\quad \Rightarrow G_1 \cong G_3$.

Remark. Clearly (by 22.4) isomorphic finite groups must have the same order. But the converse is not true, e.g. a four-group and a cyclic group of order 4 are non-isomorphic groups of order 4.

Worked example. Prove that two finite cyclic groups of the same order are isomorphic to each other.

Solution. Let G, G_1 be finite cyclic groups, both of order k.

Let a be a generator of G and b a generator of G_1. Then (by 39.2) a and b both have period k. Moreover (by 38.4) $1, a, a^2, \ldots, a^{k-1}$ is a complete repetition-free list of the elements of G, and $1, b, b^2, \ldots, b^{k-1}$ is a complete repetition-free list of the elements of G_1.

An obvious bijection from G to G_1 is the mapping θ specified by

$$\theta(a^n) = b^n \quad\quad (n = 0, 1, 2, \ldots, k-1)$$

We shall prove that θ is an isomorphism.

As a preliminary, we prove that the equation $\theta(a^n) = b^n$ holds for all integral values of n (not just those in the range 0 to $k-1$). For this purpose, let n be an arbitrary integer, and let r be its principal remainder on division by k. Then, by 38.4(i), $a^n = a^r$ and $b^n = b^r$. Hence

$\theta(a^n) = \theta(a^r)$
$\quad\quad\quad = b^r \quad\quad$ (by definition of θ, since $0 \leqslant r \leqslant k-1$)
$\quad\quad\quad = b^n$

Now let x, y be arbitrary elements of G. Since $G = \langle a \rangle$, $x = a^s$ and $y = a^t$ for some integers s, t, and hence

$$\theta(xy) = \theta(a^s . a^t) = \theta(a^{s+t}) = b^{s+t} = b^s . b^t = \theta(a^s)\theta(a^t) = \theta(x)\theta(y).$$

Since x, y are arbitrary and θ is bijective, it follows that θ is an isomorphism from G to G_1.

Hence $G \cong G_1$, which proves the result.

47. Homomorphisms and their elementary properties

Definition. Let G, G_1 be groups. A mapping $\theta : G \to G_1$ (whether bijective or not) is called a **homomorphism** from G to G_1 iff $\theta(xy) = \theta(x)\theta(y)$ for all x, $y \in G$.

Remarks. (*a*) Obviously "isomorphism" is synonymous with "bijective homomorphism". An illustration of a non-bijective homomorphism is the mapping from $GL(n, \mathbb{R})$ to $\mathbb{R} - \{0\}$ given by $X \mapsto \det X$, where $n \geqslant 2$. This mapping det is a homomorphism because $\det(XY) = \det X \times \det Y$ for all X, $Y \in GL(n, \mathbb{R})$; and it is not injective (and therefore not bijective) because, for example, there is certainly more than one member of $GL(n, \mathbb{R})$ with determinant equal to 1.

(*b*) If θ is a homomorphism from the group G to the group G_1, then, for any elements $x_1, x_2, \ldots, x_n \in G$,

$$\theta(x_1 x_2 \ldots x_n) = \theta(x_1)\theta(x_2)\ldots\theta(x_n);$$

and, in particular, for every $x \in G$ and every $n \in \mathbb{N}$,

$$\theta(x^n) = \{\theta(x)\}^n$$

These assertions are easily proved by repeated application of the definition of a homomorphism. E.g. in the case $n = 3$ we have

$$\theta(x_1 x_2 x_3) = \theta(x_1 x_2)\theta(x_3) = \theta(x_1)\theta(x_2)\theta(x_3).$$

In the following discussion and theorems up to 47.8 (inclusive), θ will denote an arbitrary homomorphism from G to G_1 (G, G_1 being groups). For clarity, we shall denote the identities of G, G_1 by 1, 1_1, respectively.

47.1 $\theta(1) = 1_1$.

Proof. $\theta(1) = \theta(1 \cdot 1)$
$\qquad\quad = \theta(1)\theta(1)$ (since θ is a homomorphism);
i.e. $\theta(1) \cdot 1_1 = \theta(1)\theta(1)$.
Hence, by cancellation in G_1, $1_1 = \theta(1)$.

47.2 $\forall x \in G$, $\theta(x^{-1}) = \{\theta(x)\}^{-1}$.

Proof. Let x be an arbitrary element of G. We have

$$\theta(x)\theta(x^{-1}) = \theta(xx^{-1}) \qquad \text{(since } \theta \text{ is a homomorphism)}$$
$$= \theta(1)$$
$$= 1_1 \qquad \text{(by 47.1)}$$

This shows (by 35.3) that $\theta(x^{-1})$ is the inverse of $\theta(x)$ in the group G_1, i.e. $\theta(x^{-1}) = \{\theta(x)\}^{-1}$.

47.3 If H is a subgroup of G, then $\theta(H)$ is a subgroup of G_1.

Proof. Let H be a subgroup of G.

(1) Since $H \neq 0$, $\theta(H) \neq 0$.

(2) Let x, y be arbitrary elements of $\theta(H)$. Then $x = \theta(h_1)$ and $y = \theta(h_2)$ for some h_1, $h_2 \in H$. Hence, using 47.2, we have

$$x^{-1}y = \{\theta(h_1)\}^{-1}\theta(h_2) = \theta(h_1^{-1})\theta(h_2) = \theta(h_1^{-1}h_2).$$

Since H is a subgroup, $h_1^{-1}h_2 \in H$, and hence $x^{-1}y \in \theta(H)$.

This proves that x, $y \in \theta(H) \Rightarrow x^{-1}y \in \theta(H)$.

It follows (by 36.4) that $\theta(H)$ is a subgroup of G_1.

An important special case of 47.3 (the case $H = G$) is:

47.4 $\operatorname{im} \theta$ is a subgroup of G_1.

The above theorems, it should be noted, are all variations on the theme "homomorphisms respect features of groups". By combining 47.2 with the remark (b) on the definition of a homomorphism, it is easy to prove further that:

47.5 For every $x \in G$ and for every integer n, $\theta(x^n) = \{\theta(x)\}^n$.

We use words for special kinds of homomorphisms as follows:

(a) an injective homomorphism is called a **monomorphism**;

(b) a surjective homomorphism is called an **epimorphism**;

(c) (as already noted) a bijective homomorphism is called an **isomorphism**.

If θ is a monomorphism, then the restriction of θ obtained by reducing the codomain to $\operatorname{im} \theta$ (a group, by 47.4) is a bijection (and still a homomorphism), i.e. it is an isomorphism from G to $\operatorname{im} \theta$. Therefore:

47.6 If θ is a monomorphism with domain G, then $\operatorname{im} \theta \cong G$.

Kernel of a homomorphism

As before, θ denotes an arbitrary homomorphism from G to G_1.

Definition. The **kernel** of θ (denoted by $\ker \theta$) is the subset

$$\{x \in G : \theta(x) = 1_1\}$$

of the domain G of θ.

Straight away we show that:

47.7 $\ker \theta$ is a subgroup of G.

Proof. (1) Since (by 47.1) $\theta(1) = 1_1$, $1 \in \ker \theta$. So $\ker \theta \neq 0$.

(2) Let x, $y \in \ker \theta$, so that $\theta(x) = \theta(y) = 1_1$. Then

$$\begin{aligned}
\theta(x^{-1}y) &= \theta(x^{-1})\theta(y) &&\text{(since } \theta \text{ is a homomorphism)} \\
&= \{\theta(x)\}^{-1}\theta(y) &&\text{(by 47.2)} \\
&= 1_1^{-1}1_1 \\
&= 1_1
\end{aligned}$$

and so $x^{-1}y \in \ker \theta$.

This shows that $x, y \in \ker \theta \Rightarrow x^{-1}y \in \ker \theta$.

It follows (by 36.4) that $\ker \theta$ is a subgroup of G.

Illustration. Consider, as before, the homomorphism $\det : GL(n, \mathbb{R}) \to \mathbb{R} - \{0\}$. The kernel of this homomorphism is the subgroup of $GL(n, \mathbb{R})$ consisting of all the matrices in $GL(n, \mathbb{R})$ with determinant 1.

The kernel of a homomorphism turns out to be a very useful object to consider. For example, as the next theorem shows, a knowledge of the kernel of a homomorphism tells us immediately whether the homomorphism is injective.

47.8 Let $\theta : G \to G_1$ be a homomorphism. Then θ is a monomorphism if and only if $\ker \theta$ is trivial.

Proof. (i) Suppose that θ is a monomorphism. By 47.1, $1 \in \ker \theta$. If $\ker \theta$ contained a second element $t \neq 1$, we would have $\theta(t) = \theta(1)$ $(=1_1)$—a contradiction since θ is injective. Therefore $\ker \theta$ must consist of 1 alone.

Thus θ is a monomorphism $\Rightarrow \ker \theta = \{1\}$.

(ii) Conversely, suppose that $\ker \theta = \{1\}$. Then, for $x, y \in G$,

$$\begin{aligned}
\theta(x) = \theta(y) &\Rightarrow \theta(x)\{\theta(y)\}^{-1} = \theta(y)\{\theta(y)\}^{-1} \\
&\Rightarrow \theta(xy^{-1}) = 1_1 \qquad \text{(by 47.2)} \\
&\Rightarrow xy^{-1} \in \ker \theta \qquad \text{(by definition of } \ker \theta) \\
&\Rightarrow xy^{-1} = 1 \qquad \text{(since } \ker \theta = \{1\}) \\
&\Rightarrow x = y
\end{aligned}$$

so that θ is injective.

Thus $\ker \theta = \{1\} \Rightarrow \theta$ is a monomorphism.

Homomorphisms where the group operations need not be multiplication
Suppose that $(G, \circ)$ and $(G_1, *)$ are groups, with binary operations $\circ, *$, respectively. Then a homomorphism from G to G_1 means a mapping $\theta : G \to G_1$ such that

$$\forall x, y \in G, \ \theta(x \circ y) = \theta(x) * \theta(y)$$

All our results on homomorphisms extend to this more general situation with suitable adjustments in notation.

Illustration. Let G be the multiplicative group of all positive real numbers, and let G_1 be the additive group of all real numbers. Then the logarithm function, $\log : G \to G_1$, given by $x \mapsto \log x$, is (for any definite choice of base of logarithms) a homomorphism from G to G_1, since

$$\log(xy) = \log x + \log y \quad \text{for all} \quad x, y \in G.$$

Indeed log is actually an isomorphism from G to G_1.

Similarly the exponential function is (in the real or complex context) a homomorphism, since

$$\exp(x+y) = (\exp x).(\exp y) \quad \text{for all relevant } x, y.$$

The fact that the logarithm function is an isomorphism from the multiplicative group of positive numbers (a system in which calculation is relatively difficult) to the additive group of real numbers (a system in which calculation is relatively easy) explains in a sophisticated way the arithmetical usefulness of logarithms.

Cayley's theorem
The final result in this section (a theorem associated with the nineteenth-century British mathematician Cayley) gives a striking hint of the value, for group theory as a whole, of the study of groups of permutations.

47.9 Every group is isomorphic to a group of permutations.

Proof. Let G be an arbitrary group.

Let S be the symmetric group on the set G, i.e. the group of all permutations of G. [We shall show that G is isomorphic to a subgroup of S.] By 35.8, we know that $\lambda_x \in S$ for each $x \in G$ (where λ_x denotes left multiplication by x). So we can define a mapping $\theta : G \to S$ by $\theta(x) = \lambda_x$.

We now note that, for all $x, y, g \in G$,

$$(\lambda_x \circ \lambda_y)(g) = \lambda_x(\lambda_y(g)) = \lambda_x(yg) = x(yg) = (xy)g = \lambda_{xy}(g)$$

This proves that, for all $x, y \in G$,

$$\lambda_x \circ \lambda_y = \lambda_{xy}, \quad \text{i.e.} \quad \theta(x) \circ \theta(y) = \theta(xy)$$

and so θ is a homomorphism.

Consider now $\ker \theta$. For $x \in G$,

$$x \in \ker \theta \Rightarrow \theta(x) = \text{identity of the group } S$$
$$\Rightarrow \lambda_x = i_G$$
$$\Rightarrow \lambda_x(1) = i_G(1)$$
$$\Rightarrow x = 1 \quad \text{(by the definitions of } \lambda_x \text{ and } i_G\text{)}.$$

Therefore $\ker \theta$ can contain no element other than 1, and so, since certainly $1 \in \ker \theta$, $\ker \theta = \{1\}$.

By 47.8, it follows that θ is a monomorphism. Hence, by 47.6, $G \cong \operatorname{im} \theta$. Since $\operatorname{im} \theta$ is a subgroup of S and, therefore, $\operatorname{im} \theta$ is a group of permutations, the result follows.

48. Conjugacy

Throughout this section, G will denote a group.

For $a, x \in G$, the **conjugate of** a **by** x means the element xax^{-1} of G. And, for given $a \in G$, each element of G expressible as xax^{-1} for some $x \in G$ is called a **conjugate** of a (in G).

Conjugacy (in G) is the relation $\sim$ on G defined by

$$a \sim b \Leftrightarrow b \text{ is a conjugate of } a,$$

i.e.

$$a \sim b \Leftrightarrow \exists x \in G \text{ s.t. } b = xax^{-1}$$

Leaving the proof as an exercise, we record that:

48.1 Conjugacy is an equivalence relation on G.

Because of this fact, G can be partitioned into $\sim$-classes, which we call **conjugacy classes**. Thus the conjugacy class of a (where $a \in G$) means the set of all conjugates of a in G: we denote this subset of G by $\mathrm{cl}(a)$.

To obtain information about $|\mathrm{cl}(a)|$ when G is finite, we must consider the question: when are two conjugates gag^{-1}, hah^{-1} of a equal and when are they different $(g, h \in G)$? In fact, this question has already been answered—in worked example 2 in §42: from the result proved there, we know that the conjugates of a by two elements of G are

$$\begin{cases} \text{the same if the elements belong to the same left coset of } C(a), \\ \text{different if the elements belong to different left cosets of } C(a). \end{cases}$$

It follows that:

48.2 In a finite group G, for each $a \in G$, the number of different conjugates of a in G equals the number of different left cosets of $C(a)$ in G, i.e.

$$|\mathrm{cl}(a)| = |G : C(a)|$$

and consequently $|\mathrm{cl}(a)|$ is a divisor of $|G|$.

(The final assertion follows by Lagrange's theorem (43.1).)

Note that $\mathrm{cl}(a)$ may consist of a alone $(a \in G)$. In fact:

$$\mathrm{cl}(a) = \{a\} \Leftrightarrow \text{every conjugate of } a \text{ is equal to } a$$
$$\Leftrightarrow \forall x \in G, \; xax^{-1} = a$$
$$\Leftrightarrow \forall x \in G, \; xa = ax$$
$$\Leftrightarrow a \in Z(G), \text{ the centre of } G \text{ (see §37(2))}.$$

Therefore:

48.3 If Z is the centre of G, the elements of Z form single-membered conjugacy classes, and the elements of $G - Z$ belong to multi-membered conjugacy classes.

This situation is illustrated in the diagram below, which portrays the partition of G into conjugacy classes: the centre Z divides into single-membered classes (one of which is $\{1\}$), while $G - Z$ divides into multi-membered classes.

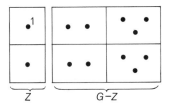

We end this section with a lemma which is of great significance because it can be used to open up an important seam of results about finite groups whose order is a power of a prime, e.g. later in the chapter, we shall use this lemma in our proof that a group whose order is the square of a prime must be abelian.

48.4 Let G be a group of order p^r, where p is a prime and $r \in \mathbb{N}$. Then the centre of G is non-trivial.

Proof. The result is certainly true if G is abelian: for then (cf. 37.3) the centre of G is the whole of G.

So henceforth we consider the case G non-abelian. Let Z be the centre of G, and let $C_1, C_2, \ldots, C_t$ be the multi-membered conjugacy classes of G, i.e. the conjugacy classes into which the non-empty set $G - Z$ partitions (see diagram). Clearly, then, $|G - Z| = \sum_{j=1}^{t} |C_j|$. As $|Z| = |G| - |G - Z|$, we deduce that

$$|Z| = |G| - \sum_{j=1}^{t} |C_j| \qquad (\alpha)$$

Now each $|C_j|$ is, by 48.2, a factor of $|G|$, i.e. of p^r, and is greater than 1. So each $|C_j|$ is p or p^2 or p^3 or $\ldots$, and in particular each $|C_j|$ is divisible by p. Therefore, in the equation (α) every term of the right-hand side is divisible by p, and hence $|Z|$ is divisible by p. Hence $|Z| \neq 1$.

The result now follows.

49. Normal subgroups

Throughout this section, G will denote a group and H a subgroup of G.

As we saw in earlier discussion of cosets, for $x \in G$, the left coset xH need not equal the right coset Hx. This is the background for the definition of a normal subgroup.

Definition. Given that H is a subgroup of G, we say that H is a **normal** subgroup of G (or that H is normal in G) iff $xH = Hx$ for all $x \in G$. The standard notation for "H is a normal subgroup of G" is

$$H \lhd G$$

When H is a normal subgroup of G, it is clear from the definition that we may forget the distinction between left and right cosets of H and speak simply of "cosets of H". In § 50 it will be shown that the cosets of a normal subgroup of G form a group with respect to multiplication of subsets of G—a fact without any parallel in the case of a non-normal subgroup. Meanwhile, we consider theorems that provide tests for normality and show that subgroups of certain kinds are normal.

49.1 In every group G, the subgroups G and $\{1\}$ are normal.

Proof. Clearly, $\forall x \in G$, $xG = Gx = G$ and $x\{1\} = \{1\}x = \{x\}$.

49.2 A subgroup of index 2 is always normal.

Proof. Let H be a subgroup with index 2 in the group G.

Then there are just two left cosets of H in G. One of these is H, and, since the left cosets form a partition of G, the other must be $G - H$. But also, by 43.2, there are just 2 right cosets of H in G: again one of them is H and hence the other is $G - H$. It follows, by 42.5, that, for every $x \in G$,

$$xH = Hx = \begin{cases} H & \text{if } x \in H \\ G - H & \text{if } x \notin H \end{cases}$$

Therefore $H \lhd G$, which proves the result.

The next theorem gives a very useful test for normality.

49.3 Let H be a subgroup of the group G. Then $H \lhd G$ iff the condition

$$h \in H \text{ and } x \in G \Rightarrow xhx^{-1} \in H$$

holds (i.e. iff every conjugate of an element of H by an element of G belongs to the subgroup H).

Proof. (i) Suppose that $H \lhd G$. Then

$$\begin{aligned} h \in H \text{ and } x \in G &\Rightarrow xh \in xH \\ &\Rightarrow xh \in Hx \qquad \text{(by the meaning of } H \lhd G) \\ &\Rightarrow xhx^{-1} \in H \qquad \text{(cf. 42.6).} \end{aligned}$$

Thus $H \lhd G$ implies that the condition holds.

(ii) Conversely, suppose that the condition holds.

Let x be an arbitrary element of G. For $t \in G$, we have

$$t \in xH \Rightarrow t^{-1}x \in H \qquad \text{(cf. 42.6)}$$
$$\Rightarrow t(t^{-1}x)t^{-1} \in H \qquad \text{(by the condition)}$$
$$\Rightarrow xt^{-1} \in H$$
$$\Rightarrow t \in Hx \qquad \text{(cf. 42.6).}$$

and conversely

$$t \in Hx \Rightarrow xt^{-1} \in H$$
$$\Rightarrow t^{-1}(xt^{-1})(t^{-1})^{-1} \in H \qquad \text{(by the condition)}$$
$$\Rightarrow t^{-1}x \in H$$
$$\Rightarrow t \in xH.$$

Hence $t \in xH \Leftrightarrow t \in Hx$ $(t \in G)$; and so $xH = Hx$.

This shows that $\forall x \in G$, $xH = Hx$, i.e. that $H \triangleleft G$.

Thus the condition implies $H \triangleleft G$.

Proof of the stated result is now complete.

We apply 49.3 in proving the next two results.

49.4 The kernel of a homomorphism with G as domain is a normal subgroup of G.

Proof. Let $\theta : G \rightarrow G_1$ be a homomorphism (G, G_1 being groups). By 47.7, $\ker \theta$ is a subgroup of G.

Suppose that $k \in \ker \theta$ and $x \in G$. Then

$$\theta(xkx^{-1}) = \theta(x)\theta(k)\{\theta(x)\}^{-1}$$
$$= \theta(x)1_1\{\theta(x)\}^{-1}, \quad \text{where } 1_1 \text{ is identity of } G_1 \text{ (since } k \in \ker \theta)$$
$$= \theta(x)\{\theta(x)\}^{-1}$$
$$= 1_1$$

so that $xkx^{-1} \in \ker \theta$.

Thus $k \in \ker \theta$ and $x \in G \Rightarrow xkx^{-1} \in \ker \theta$.

So, by 49.3, $\ker \theta \triangleleft G$, which proves the result.

49.5 Every subgroup of G that is contained in the centre Z of G (including Z itself) is normal in G.

Proof. Let H be a subgroup of G such that $H \subseteq Z$, where Z is the centre of G. Then, for every $h \in H$ and every $x \in G$,

$$xhx^{-1} = hxx^{-1} \qquad (h \text{ commutes with } x \text{ since } h \in H \text{ and } H \subseteq Z)$$
$$= h$$

so that $xhx^{-1} \in H$.

Thus $h \in H$ and $x \in G \Rightarrow xhx^{-1} \in H$.

So, by 49.3, $H \lhd G$, which proves the result.

A particular case of 49.5 is the following result, which is, in fact, immediately obvious from the original definition of normality.

49.6 In an abelian group, every subgroup is a normal subgroup.

50. Quotient groups

Once again, G will, throughout the section, denote a group.

We now set off to show that if $K \lhd G$, the cosets of K in G form a group with respect to multiplication of subsets of G. We shall require the result of one of the exercises on chapter 7, viz:

50.1 For every subgroup H of G, $HH = H$.

Henceforth suppose that $K \lhd G$. For $x, y \in G$, we have:

$$(xK)(yK) = xKyK \qquad \text{(brackets can be dropped because subset multiplication is associative (41.1))}$$
$$= xyKK \qquad (Ky = yK \text{ because } K \lhd G)$$
$$= xyK \qquad \text{(by 50.1)}.$$

Hence:

50.2 When $K \lhd G$, the set of cosets of K in G is closed under multiplication (of subsets of G); and we have the simple rule

$$(xK)(yK) = xyK \qquad (x, y \in G).$$

Note: the "simple rule" says that, for $x, y \in G$, $C_x C_y = C_{xy}$, where C_t denotes the coset of K to which t belongs.

We now build on 50.2 to produce the main result.

50.3 When $K \lhd G$, the set of cosets of K in G is a group (with respect to multiplication of subsets of G).

Proof. Let S be the set of cosets of K in G $(K \lhd G)$. We check the group axioms in turn for S.

G0: Holds for S by 50.2.

G1: Holds for S since the multiplication of subsets of G is associative.

G2: The coset K itself acts as an identity in S: for, $\forall x \in G$,

$$K(xK) = (1K)(xK) = (1x)K = xK, \text{ and } (xK)K = (xK)(1K) = (x1)K = xK.$$

Thus G2 holds for S.

G3: Each coset xK $(x \in G)$ has an inverse in S (viz. $x^{-1}K$), since

$$(xK)(x^{-1}K) = (xx^{-1})K = 1K = K, \text{ and } (x^{-1}K)(xK) = (x^{-1}x)K = 1K = K$$

and since K has been recognized as identity element. Thus G3 holds for S. It now follows that S is a group, as asserted.

Nomenclature. When $K \lhd G$, the group formed by the cosets of K in G is called the **quotient group** (or **factor group**) of G modulo K. It is denoted by G/K. Note explicitly:

50.4 When $K \lhd G$,

    (i) the identity of the group G/K is K,

    (ii) the inverse of the element xK in G/K is $x^{-1}K$ $(x \in G)$,

    (iii) if $|G:K|$ is finite, then $|G/K| = |G:K|$.

    ((i) and (ii) emerged in the course of the proof of 50.3; and (iii) is obvious).

It should also be noted that the simple rule for multiplying elements of G/K can, through repeated application in conjunction with 50.4, be extended to longer products, powers, etc. For example,

$$(x_1 K)(x_2 K)\ldots(x_n K) = x_1 x_2 \ldots x_n K \qquad (x_1, x_2, \ldots, x_n \in G)$$

$$(xK)(yK)^{-1}(zK) = xy^{-1}zK \qquad (x, y, z \in G)$$

$$(xK)^n = x^n K \qquad (x \in G, n \in \mathbb{Z}).$$

Such observations are associated with the fact that:

50.5 When $K \lhd G$, the mapping $v : G \to G/K$ given by $x \mapsto xK$ is an epimorphism with kernel K.

Proof. When $K \lhd G$, we have, for every $x, y \in G$,

$$\begin{aligned} v(xy) &= xyK & \text{(by definition of } v) \\ &= (xK)(yK) & \text{(by rule of multiplication in } G/K) \\ &= v(x)v(y) \end{aligned}$$

Therefore v is a homomorphism. Moreover, v is surjective, since the arbitrary element xK of G/K $(x \in G)$ is equal to $v(x)$. So v is an epimorphism. Finally, for $x \in G$,

$$\begin{aligned} x \in \ker v &\Leftrightarrow v(x) = \text{identity of } G/K \\ &\Leftrightarrow xK = K \\ &\Leftrightarrow x \in K \qquad \text{(by 42.6)}, \end{aligned}$$

so that $\ker v = K$.

Remarks. (*a*) The epimorphism v introduced in 50.5 is called the **natural epimorphism** from G onto G/K.

(*b*) In 49.4 it was proved that the kernel of a homomorphism with the group G as domain is a normal subgroup of G. From 50.5 it emerges that, conversely, every normal subgroup of the group G is the kernel of at least one homomorphism with G as domain.

(*c*) We have shown that, when $K \lhd G$, the set of (left) cosets of K in G can be made into a group with the rule of multiplication

$$(\text{coset of } x) . (\text{coset of } y) = \text{coset of } xy \qquad (x, y \in G)$$

It is natural to ask: is it ever possible, when H is a non-normal subgroup of G, to make the set of left cosets of H in G into a group with the rule of multiplication

$$(\text{left coset of } x) . (\text{left coset of } y) = \text{left coset of } xy \qquad (x, y \in G)?$$

The answer is no: for, if it were possible, the mapping from G to the group formed by the cosets given by $x \mapsto (\text{left coset of } x)$ would, fairly obviously, be a homomorphism with H as kernel — an impossibility unless $H \lhd G$.

51. The quotient group G/Z

Let G be a group, and let Z be its centre. By 49.5, $Z \lhd G$, and so the quotient group G/Z exists. An interesting and useful lemma about this quotient group is:

51.1 The group G/Z, where Z is the centre of the group G, cannot be non-trivial cyclic.

Proof. Suppose that G/Z is cyclic.

Then $G/Z = \langle \tau \rangle$ for some $\tau \in G/Z$, i.e. $G/Z = \langle tZ \rangle$ for some $t \in G$. So each element of G/Z, i.e. each coset of Z in G, is equal to $(tZ)^i$, i.e. to t^iZ, for some integer i.

Let x, y be arbitrary elements of G. Suppose that the cosets of Z to which they belong are t^mZ, t^nZ, respectively ($m, n \in \mathbb{Z}$). Then $x = t^mz_1$ and $y = t^nz_2$ for some $z_1, z_2 \in Z$. Hence

$$
\begin{aligned}
xy &= t^mz_1t^nz_2 \\
&= t^mt^nz_1z_2 \qquad (z_1 \text{ commutes with } t^n \text{ since } z_1 \in Z) \\
&= t^{m+n}z_1z_2
\end{aligned}
$$

and $\quad$
$$
\begin{aligned}
yx &= t^nz_2t^mz_1 \\
&= t^nt^mz_1z_2 \qquad (z_2 \text{ commutes with } t^mz_1 \text{ since } z_2 \in Z) \\
&= t^{m+n}z_1z_2
\end{aligned}
$$

and hence $xy = yx$.

Since x, y are arbitrary, this proves that G is abelian, i.e. that Z is the whole of G, i.e. that G/Z is trivial.

It follows that G/Z cannot be cyclic and non-trivial.

As a corollary, we now prove:

51.2 If G is a group of order p^2, where p is prime, then G is abelian.

Proof. Let G be a group of order p^2, where p is prime.

Let Z be the centre of G. By Lagrange's theorem, $|Z|\,|\,|G|$ and so $|Z|$ is 1 or p or p^2.

But $|Z| \neq 1$, by 48.4; and $|Z| \neq p$ since

$$|Z| = p \Rightarrow |G/Z| = |G|/|Z| = p^2/p = p$$
$$\Rightarrow G/Z \text{ is non-trivial and cyclic (by 44.2)}$$

—an impossibility, by 51.1.

Therefore $|Z| = p^2$, i.e. Z is the whole of G, i.e. G is abelian. This proves the result.

52. The first isomorphism theorem

We saw in §50 that, when K is a normal subgroup of the group G, there is a homomorphism v such that

$$\text{im } v = G/K = G/\text{ker } v.$$

The following result, known as *the first isomorphism theorem*, may be seen as a generalization of what we have just mentioned.

52.1 Let $\theta : G \to G_1$ be a homomorphism, G, G_1 being groups. Then

$$\text{im } \theta \cong G/\text{ker } \theta.$$

Proof. Let $K = \text{ker } \theta$. (The group G/K exists, by virtue of 49.4.)

The proof hinges on the possibility of defining a mapping $\phi : G/K \to G_1$ consistently by

$$\phi(xK) = \theta(x) \qquad (x \in G) \tag{ω}$$

The problem of consistency (encountered before in §18) arises here because each member of G/K is likely to have several labels of the form tK $(t \in G)$, and so one must investigate the question: if the element ξ of G/K is equal both to xK and to yK $(x, y \in G)$, do we necessarily come, in using the equation (ω) to determine $\phi(\xi)$, to the same conclusion when using the label xK for ξ as when using the label yK, i.e. is $\theta(x)$ necessarily the same as $\theta(y)$? The answer is yes, because, for $x, y \in G$,

$$\begin{aligned}
xK = yK &\Leftrightarrow x^{-1}y \in K \ (= \text{ker } \theta) \qquad \text{(by 42.6)} \\
&\Leftrightarrow \theta(x^{-1}y) = 1_1 \qquad (1_1 \text{ being the identity of } G_1) \\
&\Leftrightarrow \{\theta(x)\}^{-1}\theta(y) = 1_1 \qquad \text{(since } \theta \text{ is a homomorphism)} \\
&\Leftrightarrow \theta(x) = \theta(y)
\end{aligned}$$

So we may proceed knowing that the equation (ω) consistently defines a mapping ϕ from G/K to G_1. Moreover, since the above analysis shows that, for

$x, y \in G$,

$$\theta(x) = \theta(y) \Rightarrow xK = yK$$

i.e.

$$\phi(xK) = \phi(yK) \Rightarrow xK = yK$$

the mapping ϕ is injective. We also observe that

$$\operatorname{im} \phi = \{\phi(xK): x \in G\} = \{\theta(x): x \in G\} = \operatorname{im} \theta.$$

We note finally that ϕ is a homomorphism, since, for all $x, y \in G$,

$$
\begin{array}{ll}
\phi(xK \cdot yK) = \phi(xyK) & \text{(by rule of multiplication in } G/K) \\
\quad = \theta(xy) & \text{(by definition of } \phi) \\
\quad = \theta(x)\theta(y) & \text{(since } \theta \text{ is a homomorphism)} \\
\quad = \phi(xK)\phi(yK) & \text{(by definition of } \phi)
\end{array}
$$

To sum up, we have proved that ϕ is a monomorphism whose image set equals $\operatorname{im} \theta$. By 47.6, it follows that

$$\operatorname{im} \theta \cong \text{domain of } \phi, \quad \text{i.e.} \quad \operatorname{im} \theta \cong G/K, \quad \text{i.e.} \quad \operatorname{im} \theta \cong G/\ker \theta.$$

The first isomorphism theorem has many uses in the more advanced development of the theory of groups. Here we can illustrate its usefulness in telling us the structure of certain quotient groups in terms of something familiar.

For example, in $GL(n, \mathbb{R})$, the subset H comprising all the matrices with determinant 1 is, we saw in an earlier section, a subgroup. And $H \lhd GL(n, \mathbb{R})$, because H is the kernel of the homomorphism $\det: GL(n, \mathbb{R}) \to \mathbb{R} - \{0\}$ given by $X \mapsto \det X$. What, we might ask, is the structure of the factor group $GL(n, \mathbb{R})/H$? Since H is the kernel of the homomorphism $\det$ and since the image set of this homomorphism is clearly the whole of $\mathbb{R} - \{0\}$, 52.1 answers our question, telling us that $GL(n, \mathbb{R})/H \cong \mathbb{R} - \{0\}$.

Another illustration arises from the homomorphism θ from the additive group $(\mathbb{R}, +)$ to the multiplicative group $\mathbb{C} - \{0\}$ given by

$$\theta(x) = \cos(2\pi x) + i \sin(2\pi x) \qquad (x \in \mathbb{R}).$$

Here the kernel is the additive group $(\mathbb{Z}, +)$ and the image set is the multiplicative group T of all complex numbers with modulus 1. Hence the additive quotient group $\mathbb{R}/\mathbb{Z}$ is isomorphic to the multiplicative group T.

Worked problem. Let θ be a homomorphism from G to G_1, G and G_1 being finite groups. Show that $|\operatorname{im} \theta|$ divides $\operatorname{hcf}(|G|, |G_1|)$.

Solution. By the first isomorphism theorem, $\operatorname{im} \theta \cong G/K$, where $K = \ker \theta$, and hence

$$|\operatorname{im} \theta| = |G/K| = |G:K|,$$

from which it follows, by Lagrange's theorem that $|\operatorname{im}\theta|\,|\,|G|$. But since $\operatorname{im}\theta$ is a subgroup of G_1 (cf. 47.4), it follows also from Lagrange's theorem that $|\operatorname{im}\theta|\,|\,|G_1|$. Hence (by 12.4) $|\operatorname{im}\theta|\,|\,\operatorname{hcf}(|G|,|G_1|)$.

EXERCISES ON CHAPTER EIGHT

1. Let X, Y be finite sets with $|X| = |Y|$, and let S_X, S_Y be the symmetric groups on X, Y, respectively. Prove that $S_X \cong S_Y$, by introducing a bijection $\alpha : X \to Y$ and considering the mapping $\theta : S_X \to S_Y$ given by $\theta(f) = \alpha \circ f \circ \alpha^{-1}$ ($f \in S_X$).

2. Let K be the multiplicative group of all matrices of the form $\begin{bmatrix} 1 & t \\ 0 & 1 \end{bmatrix}$ ($t \in \mathbb{R}$) (cf. exercise 5 on chapter 6). Prove that $(K, .) \cong (\mathbb{R}, +)$.

3. (i) Let G_1, G_2, G_3 be groups, and let $\theta : G_1 \to G_2$, $\phi : G_2 \to G_3$ be homomorphisms. Prove that $\phi \circ \theta$ is a homomorphism.

(ii) Suppose that $\theta : G \to G_1$ is an isomorphism (G, G_1 being groups). Prove that θ^{-1} is an isomorphism from G_1 to G.

[*Note*: This exercise fills in some of the detail omitted from the proof of 46.1.]

4. Let $\theta : G \to G_1$ be a homomorphism (G, G_1 being groups), and let x be an element of G.

(i) Prove that $\theta(\langle x \rangle) = \langle \theta(x) \rangle$.

(ii) Suppose that x has finite period k. Prove that $\theta(x)$ has finite period dividing k and that, if θ is injective, then the period of $\theta(x)$ equals k.

5. Let G_1, G_2 be groups, and let θ be the mapping from the direct product $G_1 \times G_2$ to G_1 given by $(x, y) \mapsto x$ ($x \in G_1$, $y \in G_2$). Prove that θ is an epimorphism whose kernel is isomorphic to G_2.

6. Let $\theta : G \to G_1$ be an epimorphism (G, G_1 being groups), and let Z, Z_1 be the centres of G, G_1, respectively. Prove that $\theta(Z) \subseteq Z_1$.

7. Prove that the intersection of two normal subgroups of G is also a normal subgroup of G.

8. (i) Let H be a subgroup of G. Show that $H \lhd G$ iff $xHx^{-1} = H$ for every $x \in G$. Deduce (with the aid of exercise 5 on chapter 7) that if G contains exactly one subgroup of order m ($m \in \mathbb{N}$), then that subgroup is normal in G.

(ii) Prove that the quaternion group is an example of a non-abelian group in which every subgroup is normal.

9. In the group G, let H, K be subgroups such that $K \lhd G$. Prove that the subset KH is a subgroup. Show also that if $H \lhd G$, then $KH \lhd G$.

10. Let G be the group of order 8 consisting of the matrices $\pm I, \pm A, \pm B, \pm C$, where $A = \begin{bmatrix} 1 & 0 \\ 0 & -1 \end{bmatrix}$, $B = \begin{bmatrix} 0 & -1 \\ 1 & 0 \end{bmatrix}$, $C = \begin{bmatrix} 0 & 1 \\ 1 & 0 \end{bmatrix}$, let $H = \{I, -I, A, -A\}$, and let $K = \{I, A\}$ ($= \langle A \rangle$). Leaving aside verification that G is a group, show that $H \lhd G$ and $K \lhd H$, while K is not normal in G. [*Moral*: normality is not transitive.]

11. Suppose that in a group G of order 2^n (where n is an integer $\geqslant 2$) the element x has period 2^{n-1}. Show that $x^{2^{n-2}}$ belongs to the centre of G.

12. Let G be a group. For any subset A of G, define $N(A)$ to be the subset $\{x \in G : xA = Ax\}$ of G.

(i) Prove that, for every subset A of G, $N(A)$ is a subgroup of G. Prove further that if A is a subgroup of G, then $A \lhd N(A)$ and that $N(A)$ is the largest subgroup of G in which A is normal (in the sense that, for a subgroup B, $A \lhd B \Rightarrow B \subseteq N(A)$).

(ii) Let L, M be abelian subgroups of G, and let H be the smallest subgroup that contains $L \cup M$. (That there is such a subgroup can be seen by considering the

E

intersection of all subgroups containing $L \cup M$). By considering $N(L \cap M)$, prove that $L \cap M \lhd H$.

13. Let $\theta: G \to G_1$ be a homomorphism, G and G_1 being finite groups, let $a \in G$, and suppose that $\theta(a)$ has period m in G_1. Show that the period of a in G is a divisor of $m|\ker \theta|$.

14. A subgroup H of the group G has the property that $\forall x \in G, x^2 \in H$. By considering the identity $xhx^{-1} = (xh)^2 h^{-1} (x^{-1})^2$ $(x \in G, h \in H)$, prove that $H \lhd G$. Prove further that the quotient group G/H is abelian.

15. In an infinite abelian group G, let H be the set of all elements of finite period. Prove that $H \lhd G$ and that in G/H no non-identity element has finite period.

16. For elements x, y in a group G, the *commutator* of x, y (denoted by $[x, y]$) is defined to be the element $x^{-1} y^{-1} xy$. Prove that x, y commute iff $[x, y] = 1$.
Let H, K be normal subgroups of the group G such that $H \cap K = \{1\}$. By considering the commutator $[h, k]$, where h, k are arbitrary elements of H, K, respectively, prove that each element of H commutes with every element of K.

17. For a normal subgroup N of the group G, prove that, for $x, y \in G$, $[xN, yN] = [x, y]N$, and deduce that G/N is abelian iff $[x, y] \in N$ for all $x, y \in G$.
Deduce that, for normal subgroups H, K of the group G,

$$G/H \text{ and } G/K \text{ both abelian} \Rightarrow G/(H \cap K) \text{ is abelian.}$$

If H, K are normal subgroups of the group G such that both G/H and G/K are cyclic, need it be true that $G/(H \cap K)$ is also cyclic?

18. Let G be a group and Z its centre. Show that if xZ and yZ (where $x, y \in G$) are inverses of each other in the group G/Z, then x and y commute with each other in G.

19. Give an example of a pair of groups G_1 and G_2 containing normal subgroups K_1 and K_2, respectively, such that $K_1 \cong K_2$ and $G_1/K_1 \cong G_2/K_2$ but G_1 and G_2 are not isomorphic to each other. (*Moral*: the structures of K and of G/K do not determine the structure of G.)

20. Let G be a finite group with the property that $x^2 = 1$ for all $x \in G$. Prove that $|G|$ must be a power of 2.

21. (i) Let H be a subgroup of index 2 in the group G. Prove that $\forall x \in G, x^2 \in H$.
(ii) Is it true that if H is a subgroup of index 3 in G, then x^3 must belong to H for every $x \in G$?

22. (i) Show that if x is an element of the group G and $x \notin Z(G)$, then $Z(G) \subset C(x)$ (strict containment).
(ii) Use part (i) in conjunction with 48.4 to prove (without mentioning G/Z) that a group of order p^2, where p is prime, must be abelian.

23. Let H be a subgroup of the group G. Show that $H \lhd G$ iff H is expressible as the union of some entire conjugacy classes of G.

24. Let G be a group of order p^r, where p is prime and $r \in \mathbb{N}$, and let N be a non-trivial normal subgroup of G. Use exercise 23 to prove that $N \cap Z(G)$ is non-trivial. Deduce that if $|N| = p$, then $N \subseteq Z(G)$.

25. Let G be a non-abelian group of order p^3, where p is prime.
(i) Show that $|Z(G)| = p$.
(ii) Using exercise 22(i), deduce that the number of conjugacy classes in G is $p^2 + p - 1$.

(iii) Show also, using exercise 24, that G contains just one normal subgroup of order p.

(iv) Prove finally that $x^p \in Z(G)$ for every $x \in G$.

26. As a generalization of 51.1, show that if H and N are subgroups of a group G such that $N \subseteq H$, $N \subseteq Z(G)$ and H/N is cyclic, then H is abelian.

27. Let G be a non-abelian group of order pq, where p, q are distinct primes. (i) Prove that $Z(G)$ is trivial. (ii) Noting that (by exercise 15 on chapter 6) all elements in a conjugacy class have the same period, and using the result of exercise 16 on chapter 7, prove that the number of elements of period p in G is a multiple of q. (This number is also a multiple of $p - 1$, by exercise 21 on chapter 7.)

28. Using the last exercise and exercise 22 on chapter 7, prove that every group of order 15 is cyclic.

29. Let H be a subgroup of the group G. A subgroup C of G is called a *complement* of H in G iff $G = CH$ and $H \cap C = \{1\}$. Prove that if $H \triangleleft G$, then every such complement of H is isomorphic to G/H.

Show that in the quaternion group there is a normal subgroup not possessing a complement.

30. Let G be a finite group.

(i) Let θ be a homomorphism from G to a group G_1, and let H be a subgroup of G. By an application of the first isomorphism theorem, prove that $|\theta(H)|$ divides $|H|$.

(ii) Let K be a normal subgroup of G, and suppose that the order m of K and the index $|G:K|$ of K in G are relatively prime. By applying part (i) with $\theta = v$, the natural epimorphism from G to G/K, and with H an arbitrary subgroup of order m in G, prove that K is the only subgroup of order m in G.

31. Let G be a non-abelian group of order $2p$, where p is an odd prime. Prove that G contains at least one element of period p. Let a be such an element, and let $K = \langle a \rangle$. Prove that $K \triangleleft G$. Prove further that every element of $G - K$ has period 2 and that if b is any such element, then $bab = a^{-1}$.

32. This and the remaining problems in this sequence involve *automorphisms* of groups. An automorphism of a group G means an isomorphism from G to G. Prove that the set of all automorphisms of G is a group (with respect to mapping composition). We denote this group by Aut G.

33. Let G be a group. For each $g \in G$, let τ_g be the mapping from G to G defined by $\tau_g(x) = gxg^{-1}$ $(x \in G)$. Prove that, $\forall g \in G, \tau_g \in$ Aut G.

Let θ be the mapping from G to Aut G defined by $\theta(g) = \tau_g$ $(g \in G)$. Prove that θ is a homomorphism whose kernel is $Z(G)$.

Let $I = \{\tau_g : g \in G\}$. Deduce from the last part that I is a subgroup of Aut G and that $I \cong G/Z(G)$. Show also that $I \triangleleft$ Aut G.

34. Suppose that α is an automorphism of the finite group G with the property that $\forall t \in G - \{1\}$, $\alpha(t) \neq t$. Show that, for $x, y \in G$,

$$x^{-1}\alpha(x) = y^{-1}\alpha(y) \Rightarrow x = y$$

and deduce that every element of G is expressible in the form $g^{-1}\alpha(g)$ for some $g \in G$.

If $\alpha^2 = i_G$ (i.e. $\alpha \circ \alpha = i_G$), show that $\alpha(g) = g^{-1}$ for all $g \in G$ and deduce that G is abelian of odd order.

35. Let G be a cyclic group. Show that Aut G is abelian. Show further that:

(i) if G has prime order p, then $|$Aut $G| = p - 1$;

(ii) if $|G| = 8$, then Aut G is a four-group;

(iii) if G is infinite, then $|$Aut $G| = 2$.

CHAPTER NINE

THE SYMMETRIC GROUP S_n

53. Introduction

In §34 (pp. 71–2) the symmetric group S_n of degree n (n an arbitrary positive integer) was introduced. To recall the most fundamental facts, S_n is the group consisting of all permutations of the set $\{1, 2, \ldots, n\}$, the binary operation being mapping composition; and the order of S_n is $n!$

It should be appreciated that if G is any group of permutations of a set of n objects, then the natural step of labelling these objects $1, 2, \ldots, n$ makes G a group of permutations of $\{1, 2, \ldots, n\}$ and thus a subgroup of the group which we have named S_n. So the study of S_n and its subgroups covers the study of all groups of permutations of finite sets. That such a study is likely to be of central importance in the general development of finite group theory is strongly signalled by Cayley's theorem (47.9): any group is isomorphic to a group of permutations (of a finite set if the group is finite). It is, moreover, true (today as in the past) that the groups encountered in different branches of mathematics and in its applications are very often groups of permutations of finite sets—e.g. of the set of roots of a polynomial equation, or the set of vertices in a network (or "graph"), or the set of hydrogen atoms in a methane molecule, etc. Thus there are plenty of powerful reasons for embarking on a detailed study of symmetric groups, and it is to the most basic aspects of such a study that this chapter is devoted.

In §§54 and 55 we prepare the way for major advances in our insight by focusing attention on *cycles* (permutations of a specially simple kind) and by showing that every element of S_n is expressible in a standard way in terms of cycles. This enables us in §§56 and 57 to discuss periods of elements of S_n and to obtain details of how S_n partitions into conjugacy classes. The concluding sections of the chapter (§§58, 59, 60) reveal matters which might well escape the notice of the uninitiated, but turn out to be of far-reaching importance—in particular, the existence of a natural classification of permutations into two kinds, termed *even* and *odd*, and the fact that (for each $n \geqslant 2$) the even permutations in S_n form an interesting subgroup of index 2 called the *alternating group* of degree n.

Before we proceed to detailed consideration of these topics, let us clarify matters of nomenclature and notation. In discussing S_n we shall refer to members of the set $\{1, 2, \ldots, n\}$ as **objects**. So each member of S_n is a mapping whose action is to permute "the objects" in some particular way. To eliminate any risk of confusion between objects and permutations, we shall throughout the chapter use letters of the English alphabet (e.g. a, b, c) for objects, and letters of the Greek alphabet (e.g. $\alpha, \beta, \sigma, \tau$) for member elements of S_n. In particular, we shall use ι for the identity element of the group S_n (i.e. the identity mapping of the set $\{1, 2, \ldots, n\}$ to itself). Although the binary operation in S_n is mapping composition, we shall use the notation of multiplicative group theory to the extent of writing $\sigma\tau$ for $\sigma \circ \tau$ and describing this as the "product" of σ and τ ($\sigma, \tau \in S_n$).

The specification of an element of S_n by a $2 \times n$ matrix was explained at the top of p. 72, and, if necessary, that should now be revised. The student should also make sure that he can fluently obtain the product (i.e. composition) of any two given members of S_n and the inverse of a member of S_n. E.g., in S_5 let

$$\sigma = \begin{pmatrix} 1 & 2 & 3 & 4 & 5 \\ 5 & 4 & 2 & 1 & 3 \end{pmatrix}, \quad \tau = \begin{pmatrix} 1 & 2 & 3 & 4 & 5 \\ 5 & 1 & 3 & 2 & 4 \end{pmatrix}.$$

Then, as the student should verify in detail,

$$\sigma\tau = \begin{pmatrix} 1 & 2 & 3 & 4 & 5 \\ 3 & 5 & 2 & 4 & 1 \end{pmatrix}, \quad \text{and} \quad \tau^{-1} = \begin{pmatrix} 1 & 2 & 3 & 4 & 5 \\ 2 & 4 & 3 & 5 & 1 \end{pmatrix}.$$

If a is an object and $\rho \in S_n$, we say that ρ **fixes** a iff $\rho(a) = a$; and we say that ρ **moves** a iff $\rho(a) \neq a$. In the above illustration, σ moves all of the objects 1, 2, 3, 4, 5, while τ fixes 3.

54. Cycles

Suppose that $a_1, a_2, \ldots, a_r$ are distinct objects (i.e. members of $\{1, 2, 3, \ldots, n\}$), r being greater than 1. We denote by

$$(a_1\, a_2\, a_3 \ldots a_{r-1}\, a_r)$$

the element of S_n that maps a_1 to a_2, a_2 to $a_3, \ldots, a_{r-1}$ to a_r, and a_r to a_1, and fixes all other objects—i.e. the element of S_n that moves each of $a_1, a_2, \ldots, a_r$ one place round a circle as suggested by the diagram on p. 126, while fixing all other objects.

Such a permutation is called a **cycle**, of **length** r.

In S_6, for example, $(1\, 6\, 2\, 5)$ is a cycle of length 4; it is the permutation that maps 1 to 6, 6 to 2, 2 to 5, and 5 to 1, while fixing 3 and 4.

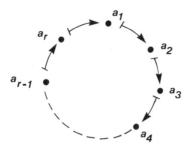

Remarks

(*a*) It is clear (especially on considering the above diagram) that

$$(a_1\,a_2\,a_3\ldots a_r), (a_2\,a_3\,a_4\ldots a_r\,a_1), (a_3\,a_4\,a_5\ldots a_r\,a_1\,a_2),\text{ etc.,}$$

all mean the same thing: that is, the choice of starting-point is immaterial when one employs the notation we have introduced for a cycle. Thus, for example, we have four possible ways of writing the cycle (1 6 2 5), namely:

$$(1\,6\,2\,5), \quad (6\,2\,5\,1), \quad (2\,5\,1\,6), \quad \text{and} \quad (5\,1\,6\,2);$$

and more generally there are r possible ways of writing a cycle of length r.

(*b*) The description of the action of a cycle α in terms of moving certain objects one place round a circle is both a natural and a helpful one. Notice that α^2 (i.e. $\alpha \circ \alpha$) will move the objects in question two places round the circle, α^3 (i.e. $\alpha \circ \alpha \circ \alpha$) will move them three places round the circle, etc. Further, α^{-1} will move all the objects one place round the circle in the reverse direction. Thus, for example, if α is (1 6 2 5), it is readily seen that α^2 is the permutation which interchanges 1 and 2 and interchanges 5 and 6, while $\alpha^{-1} = (1\,5\,2\,6)$.

(*c*) The above observations about powers of a general cycle $\alpha = (a_1\,a_2\ldots a_r)$ make it apparent that the lowest positive power of α that maps the objects $a_1, a_2, \ldots, a_r$ all the way round the circle of our earlier diagram and back to themselves is α^r. That is to say, the lowest positive power of α that equals ι is the rth power. So we can record the fact that:

54.1 The period of a cycle is equal to its length.

(*d*) A cycle of length 2 is termed a **transposition**. Such a permutation interchanges two objects while fixing all the others. A useful fact about transpositions is:

54.2 For $n \geqslant 2$, every element of S_n can be expressed as a product of transpositions.

Proof. Suppose $n \geqslant 2$, and let σ be an arbitrary member of S_n. The case $\sigma = \iota$ is easy to deal with, since $\iota = (1\,2)(1\,2)$, the product of two (equal) transpositions.

So henceforth we can confine attention to the case $\sigma \neq \iota$. In this case there are objects moved by σ. Let a be any such object. Observe that σ also moves $\sigma(a)$. [For, if $\sigma(\sigma(a))$ were equal to $\sigma(a)$, it would follow by the injectivity of σ that $\sigma(a) = a$, contrary to the fact that σ moves a.] Let τ_1 be the transposition $(a \; \sigma(a))$. Then clearly $\tau_1 \sigma$ fixes all the objects that σ fixes; and we have

$$(\tau_1 \sigma)(a) = \tau_1(\sigma(a)) = a,$$

showing that, in addition, $\tau_1 \sigma$ fixes a. Thus, given $\sigma \in S_n - \{\iota\}$, we have produced a transposition τ_1 such that $\tau_1 \sigma$ fixes more objects than σ.

If $\tau_1 \sigma \neq \iota$, we can repeat the process to produce a transposition τ_2 such that $\tau_2 \tau_1 \sigma$ fixes more objects than $\tau_1 \sigma$; and, if $\tau_2 \tau_1 \sigma \neq \iota$, we can then produce a transposition τ_3 such that $\tau_3 \tau_2 \tau_1 \sigma$ fixes yet more objects, etc. Since the number of objects is finite, it is evident that we must eventually arrive at a sequence of transpositions

$$\tau_1, \tau_2, \ldots, \tau_l$$

such that

$$\tau_l \tau_{l-1} \ldots \tau_2 \tau_1 \sigma = \iota.$$

From this equation it follows that

$$
\begin{aligned}
\sigma &= (\tau_l \tau_{l-1} \ldots \tau_2 \tau_1)^{-1} \\
&= \tau_1^{-1} \tau_2^{-1} \ldots \tau_{l-1}^{-1} \tau_l^{-1} \qquad \text{(cf. 35.7)} \\
&= \tau_1 \tau_2 \ldots \tau_{l-1} \tau_l,
\end{aligned}
$$

since, for each j, $\tau_j^2 = \iota$ (cf. 54.1) and therefore $\tau_j^{-1} = \tau_j$.

Thus σ is indeed equal to a product of transpositions, and the result is proved.

The result 54.2 is often expressed by saying that S_n is *generated* by the transpositions in it.

55. Products of disjoint cycles

We describe two cycles as **disjoint** iff the sets of objects that they move are disjoint sets. E.g., in S_8 the cycles $(1\,3\,7\,6)$ and $(2\,8\,5)$ are disjoint because the sets of objects that they move (namely, $\{1, 3, 6, 7\}$ and $\{2, 5, 8\}$) are disjoint. On the other hand, the cycles $(1\,3\,7\,6)$ and $(2\,7\,5)$ are not disjoint, because they both move the object 7.

More generally, we describe a sequence of cycles $\alpha_1, \alpha_2, \ldots, \alpha_k$ as disjoint iff they are pairwise disjoint, i.e. there is no object moved by two of them.

It turns to be important to explore, and familiarize oneself with, the action on the set of objects of a product of given disjoint cycles.

Let us begin with the case of two disjoint cycles, say $\alpha = (a_1 a_2 \ldots a_r)$ and

$\beta = (b_1 b_2 \ldots b_s)$. By appeal merely to the meaning of $\alpha\beta$ (i.e. $\alpha \circ \beta$), one soon sees that $\alpha\beta$ is the permutation which maps a_1 to a_2, a_2 to $a_3, \ldots, a_r$ to a_1, b_1 to b_2, b_2 to $b_3, \ldots, b_s$ to b_1, while fixing all other objects; that is, $\alpha\beta$ has the effect indicated in the following diagram.

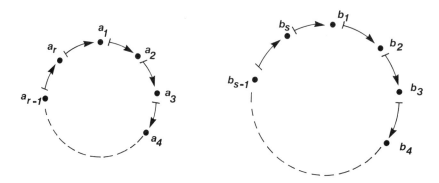

It is immediately clear that $\beta\alpha$ has precisely the same effect on all the objects, and therefore $\alpha\beta = \beta\alpha$. (They are equal mappings.) This reveals the fact that:

55.1 Disjoint cycles commute with each other.

An informal way to describe the action of the product $\alpha\beta$ of two disjoint cycles α, β as above is to say that the effect of $\alpha\beta$ is the "union" of the effects of α and β. This refers to the fact that $\alpha\beta$ has the same effect as α on the objects α moves and the same effect as β on the objects β moves—i.e. that we get a diagram indicating the action of $\alpha\beta$ by putting together a circle showing the action of the cycle α and a circle showing the action of β.

By appeal to nothing more sophisticated than the meaning of a product $\alpha_1 \alpha_2 \ldots \alpha_k$ in S_n (i.e. $\alpha_1 \circ \alpha_2 \circ \ldots \circ \alpha_k$), one can readily see that the remarks just made extend to products of any number of disjoint cycles. Informally one could say that the effect of a product $\sigma = \alpha_1 \alpha_2 \ldots \alpha_k$ of disjoint cycles $\alpha_1, \alpha_2, \ldots, \alpha_k$ is the union of the effects of $\alpha_1, \alpha_2, \ldots,$ and α_k—i.e. that one obtains a diagram indicating the action of σ by bringing together in one diagram circles showing the actions of the cycles $\alpha_1, \alpha_2, \ldots, \alpha_k$. In more formal terms, the point which has emerged is this:

55.2 If the permutation σ is the product of the disjoint cycles $\alpha_1, \alpha_2, \ldots, \alpha_k$, then σ has the same effect as α_1 on the objects α_1 moves, the same effect as α_2 on the objects α_2 moves, etc., and σ fixes the objects moved by none of $\alpha_1, \alpha_2, \ldots, \alpha_k$.

The student should realise that if a particular permutation σ is given to us in the form of a product of disjoint cycles, then because of 55.2 we can see *at a glance* what $\sigma(a)$ is for each object a. Consider for example the permutation

$$\sigma = (1\ 4)(2\ 8\ 10\ 7)(3\ 9\ 5) \qquad (\in S_{10}).$$

If we require (for instance) to know what $\sigma(7)$ is, we can see *at once* that the answer is 2 (because (cf. 55.2) σ has the same effect on 7 as the cycle in which 7 appears, namely $(2\ 8\ 10\ 7)$).

Having thought through the details that lead to 55.2, one can readily see that it generalizes to any product of permutations (cycles or other permutations) which move pairwise disjoint sets of objects. The more general result, which we shall make use of in §56, is as follows.

55.3 If the permutation σ is the product $\rho_1\rho_2\ldots\rho_k$ of permutations $\rho_1, \rho_2, \ldots, \rho_k$ which move pairwise disjoint sets of objects, then σ has the same effect as ρ_1 on the objects ρ_1 moves, the same effect as ρ_2 on the objects ρ_2 moves, etc., and σ fixes the objects moved by none of $\rho_1, \rho_2, \ldots, \rho_k$.

Our next aim is to demonstrate the immensely useful fact that any given non-identity permutation σ in S_n can be expressed as a product of disjoint cycles (in an essentially unique way). To clarify the explanation of how an arbitrary given σ may be so expressed, we shall illustrate each stage by referring to the particular permutation

$$\sigma^* = \begin{pmatrix} 1 & 2 & 3 & 4 & 5 & 6 & 7 & 8 & 9 & 10 \\ 7 & 4 & 1 & 10 & 9 & 6 & 3 & 2 & 5 & 8 \end{pmatrix} \qquad (\in S_{10}).$$

We begin by choosing an object moved by σ: we find its image under σ, then the image of that image under σ, and so on until we get back to the object we started with.

To illustrate with σ^*, let us start with the object 1. Its image under σ^* is 7; the image of 7 under σ^* is 3; and the image of 3 is 1, which brings us back to the object we started with. Thus we have found that, under σ^*,

$$1 \mapsto 7 \mapsto 3 \mapsto 1.$$

and hence there comes to our notice the cycle $(1\ 7\ 3)$, a cycle with the key property of having the same effect on the objects it moves as has the given permutation.

There is a vital detail in what has been said so far that cries out for justification. Why is it that, in the process just described and illustrated, we get back to the object we started with? Obviously we must (because the set of objects is finite) generate a repetition at some point in our chain

$$1 \mapsto 7 \mapsto \ldots;$$

but why is it that the first sign of a repetition is the reappearance of the starting-object 1? Might we not, in some other example, generate a chain such as

$$1 \mapsto \underline{4} \mapsto 8 \mapsto 3 \mapsto \underline{4} \ldots, \qquad (*)$$

with the first repeated object (4) different from the starting-object? The answer comes from the injectivity of the permutation that we are dealing with. The chain (∗) indicates that our permutation maps both 1 and 3 to 4—an impossibility for an injective mapping. In the same way, the first repetition in any chain produced in the way that has been described cannot be other than the reappearance of the starting-object.

So far, one stage of a process has been discussed, and the outcome of that stage is the production of a cycle ((1 7 3) in our illustration) with the same effect on the objects it moves as has the given permutation. To continue the process, we choose some so far unmentioned object that is moved by σ. In our illustration with σ^* let us take 2. We generate, as before, a chain of successive images under the given permutation, starting with the new object just chosen. In the case of σ^*, the chain we obtain on starting with 2 is

$$2 \mapsto 4 \mapsto 10 \mapsto 8 \mapsto 2,$$

and hence we discover another cycle, (2 4 10 8), which has the same effect as σ^* on certain of the objects. It is no fluke that this second cycle is disjoint from (1 7 3), the one found previously: for the occurrence of one of the objects from (1 7 3) in the chain starting from 2 would be an obvious contradiction of the injectivity of σ^*. (For example, we could not possibly generate a chain like

$$2 \mapsto 5 \mapsto 8 \mapsto \underline{7} \ldots$$

after finding the cycle (1 7 3) as above, since this would mean that the permutation maps both 1 and 8 to 7—a contradiction of its injectivity.)

For any given permutation σ, we continue in this way until we have exhausted all the objects moved by σ. From what has already been said, it is apparent that we shall produce a sequence of disjoint cycles, each of which has the same effect on the objects it moves as has σ. Moreover, since the process has been continued until all the objects moved by σ are exhausted, the objects fixed by σ will be precisely those not moved by any of the cycles we have produced.

In the case of σ^*, we have already found the cycles (1 7 3) and (2 4 10 8). Further continuation gives us the cycle (5 9), at which point all objects moved by σ^* are exhausted. So we end up with three cycles (1 7 3), (2 4 10 8), and (5 9), each of which has the same effect on the objects it moves as has σ^*. The effect of the product of these three cycles is (cf. 55.2 and the preceding informal remarks) the union of the effects of the cycles; and so the product has the same effect as σ^* on all nine objects that σ^* moves, while fixing the object (6) that

σ^* fixes. So it is clear (cf. the meaning of "equal mappings") that this product is equal to σ^*. We have thus succeeded in expressing σ^* as a product of disjoint cycles, namely

$$\sigma^* = (1\ 7\ 3)(2\ 4\ 10\ 8)(5\ 9);$$

and surely enough has been said to make it clear that our method can be applied to achieve the same goal with any given $\sigma \in S_n - \{\iota\}$.

By appeal once again to 55.2 an important supplementary point can quickly be brought to light. This is the fact that there is no possibility of variation in the cycles which appear when any given σ is expressed as a product of disjoint cycles: these cycles are uniquely determined by σ. This can be made clear through reference with 55.2 in mind to our illustration σ^*. Our previous uses of 55.2 have very much emphasised the fact that it tells us that the effect of a product of disjoint cycles can be inferred from the effects of the individual cycles. But of course we can equally well use 55.2 to make a deduction about each cycle in the product from a knowledge of the action of the product: each cycle must, by 55.2, have the same effect, so far as the objects it moves are concerned, as the product. So, for example, if the permutation σ^* is to be expressed as a product of disjoint cycles, it follows by 55.2 that the cycle in which 1 appears *must* map 1 to $\sigma^*(1)$, i.e. 7; equally it *must* map 7 to 3, and it *must* map 3 to 1, so that it must be (1 7 3). Similarly, the cycles (2 4 10 8) and (5 9) would have to appear, and thus a product of disjoint cycles equal to σ^* can only be

$$(1\ 7\ 3)(2\ 4\ 10\ 8)(5\ 9),$$

scope for variation being limited to our choice of starting-point in writing each cycle down and the order in which we put the three cycles. (That order can, of course, be varied at will since disjoint cycles commute (cf. 55.1).) As the argument just deployed could obviously be applied to an arbitrary σ, we have now completed justification of the following major conclusion.

55.4 Any given non-identity permutation in S_n can be expressed as a product of disjoint cycles, in a way that is unique apart from the possibility of varying the order in which the cycles appear and the starting-point in each.

Remarks

(*a*) We shall call the expression for a given permutation σ as a product of disjoint cycles the **d.c. form** of σ ("d.c." standing for "disjoint cycles").

(*b*) It has taken many sentences to describe how to obtain the d.c. form of a given permutation and to justify 55.4. However, once the process is understood, one can obtain the d.c. form of any particular permutation σ almost instantaneously! All one has to do is to write down the cycles which appear through generating chains of successive images under σ (e.g. the cycle (1 7 3) in the case of our illustration $\sigma*$).

(c) Sometimes it is convenient, in giving a permutation in d.c. form, to indicate fixed objects as cycles of length 1. Thus, an alternative notation for our permutation $\sigma*$ is

$$(1\ 7\ 3)(2\ 4\ 10\ 8)(5\ 9)(6).$$

When this convention is adopted, the identity element of S_n (which has so far been excluded from our considerations) can be expressed in d.c. form as

$$(1)(2)(3)\ldots(n).$$

(d) In several contexts it is useful to have a means of indicating how many cycles there are of each possible length in the d.c. form of a permutation σ. Such a means is provided by what we call the **cycle pattern** of σ: we say that σ has cycle pattern

$$1^{\lambda_1}.2^{\lambda_2}.3^{\lambda_3}\ldots.$$

if the d.c. form of σ contains λ_1 cycles of length 1 (i.e. σ fixes λ_1 objects), λ_2 cycles of length 2, λ_3 cycles of length 3, etc. The "1^{λ_1}" indication of the fixed objects may not be omitted if there are any fixed objects. On the other hand, in the notation $1^{\lambda_1}.2^{\lambda_2}.3^{\lambda_3}\ldots.$ we omit "j^{λ_j}" whenever λ_j is 0 (i.e. there are no cycles of length j); and we usually write "j^{λ_j}" simply as j if $\lambda_j = 1$. So, for example, with these conventions, the permutation

$$(1\ 4)(3\ 11)(5\ 8)(6\ 12\ 7\ 9)(2)(10)$$

in S_{12} has cycle pattern $1^2.2^3.4$. A cycle of length 5 in S_7 is a permutation with cycle pattern $1^2.5$; and a cycle of length 7 in S_7 has cycle pattern simply 7. The identity element of S_n has (for every n) cycle pattern 1^n.

(e) The d.c. form of a permutation being a standard and useful form of that permutation, it is important to acquire confidence in handling permutations in d.c. form (e.g. finding a product of two permutations) without feeling any desire to translate the permutations into any other form. In this connection one should note how extremely easy it is to write down the inverse or any positive power of a permutation σ given in d.c. form. Say $\sigma = \alpha_1\alpha_2\ldots\alpha_k$, where $\alpha_1, \alpha_2, \ldots, \alpha_k$ are disjoint cycles. Then, *because disjoint cycles commute,*

$$\sigma^{-1} = \alpha_1^{-1}\alpha_2^{-1}\ldots\alpha_k^{-1}$$

and

$$\sigma^m = \alpha_1^m\alpha_2^m\ldots\alpha_k^m \qquad \text{(for any } m \in \mathbb{N}\text{)};$$

and, as was explained in §54, it is a simple matter to write down the inverse or mth power of any of the cycles α_j. So, for example, returning again to our example permutation

$$\sigma* = (1\ 7\ 3)(2\ 4\ 10\ 8)(5\ 9)(6),$$

we find it a very straightforward task to write down the inverse, square and cube of $\sigma*$, namely:

$$(\sigma*)^{-1} = (1\ 3\ 7)(2\ 8\ 10\ 4)(5\ 9)(6);$$
$$(\sigma*)^2 = (1\ 3\ 7)(2\ 10)(4\ 8)(5)(6)(9);$$
$$(\sigma*)^3 = (2\ 8\ 10\ 4)(5\ 9)(1)(3)(6)(7).$$

56. Periods of elements of S_n

It has already been pointed out (see 54.1) that the period of a cycle is equal to its length. The following more general result shows how the period of any non-identity element of S_n can be deduced from its d.c. form (or indeed from its cycle pattern).

56.1 Let $\sigma \in S_n - \{\iota\}$. Suppose that σ is the product $\alpha_1\alpha_2\ldots\alpha_k$ of disjoint cycles $\alpha_1, \alpha_2, \ldots, \alpha_k$, of respective lengths $l_1, l_2, \ldots, l_k$, all $\geqslant 2$. Then the period of σ is the l.c.m. of $l_1, l_2, \ldots, l_k$ (i.e. the lowest positive integer divisible by all of $l_1, l_2, \ldots, l_k$).

Proof. Let m denote an arbitrary positive integer. As observed in remark (e) at the end of §55,

$$\sigma^m = \alpha_1^m \alpha_2^m \ldots \alpha_k^m.$$

Since $\alpha_1, \alpha_2, \ldots, \alpha_k$ are disjoint cycles, it is clear that the permutations $\alpha_1^m, \alpha_2^m, \ldots, \alpha_k^m$ move disjoint sets of objects. So if one of them (say α_j^m) moves object a, it follows (by 55.3) that σ^m also moves a. In particular, therefore,

$$\alpha_j^m \neq \iota \text{ for some } j \Rightarrow \sigma^m \neq \iota;$$

or, equivalently (cf. 4.1),

$$\sigma^m = \iota \Rightarrow \text{every } \alpha_j^m = \iota \qquad (j = 1, 2, \ldots, k).$$

The converse of this last statement is obviously true too, and so we have:

$$\sigma^m = \iota \Leftrightarrow \text{every } \alpha_j^m = \iota \, (1 \leqslant j \leqslant k)$$
$$\Leftrightarrow l_j | m \text{ for each } j \, (1 \leqslant j \leqslant k)$$

(by 38.4(ii), since each α_j has period l_j, by 54.1). Hence the lowest value of m ($\in \mathbb{N}$) for which $\sigma^m = \iota$ is the lowest positive integer divisible by all of $l_1, l_2, \ldots, l_k$, i.e. the l.c.m. of $l_1, l_2, \ldots, l_k$. This proves the result.

Note. In order to exclude unnecessary complications from the proof, mention of cycles of length 1 (i.e. fixed objects) was deliberately omitted in the above statement of 56.1. However, it is easily seen that an obvious alternative statement of 56.1, namely that the period of a permutation given in d.c. form is the l.c.m. of the lengths of the disjoint cycles appearing, is true whether we include or exclude cycles of length 1 in our calculations. Our calculations of periods will be very slightly less cluttered if we exclude such cycles.

As a specific example of the use of 56.1, consider the permutation

$$(1\ 14)(2\ 8)(3\ 15\ 7\ 6)(5\ 12\ 9\ 13\ 11\ 10)(4)$$

in S_{15}. By 56.1, the period of this permutation is l.c.m. $\{2, 2, 4, 6\}$, i.e. 12. And the same is true of every permutation of cycle pattern $1 \cdot 2^2 \cdot 4 \cdot 6$.

It is also easy now to answer a question such as: what is the smallest symmetric group that contains an element of period 10? The point is that a permutation of period 10, when expressed in d.c. form, must (by 56.1) contain at least either a cycle of length 5 and a cycle of length 2 or a cycle of length 10. Clearly this is possible iff the number of objects is at least 7, and so the answer to the question is S_7.

57. Conjugacy in S_n

The discussion of conjugacy in S_n begins with the observation that it is very easy indeed, when given permutations $\sigma, \tau \in S_n$ with σ expressed in d.c. form, to calculate the conjugate of σ by τ (i.e. the permutation $\tau\sigma\tau^{-1}$). The following result gives the details.

57.1 Let

$$\sigma = (a_1\ a_2 \ldots a_r)(b_1\ b_2 \ldots b_s)(c_1\ c_2 \ldots c_t)\ldots$$

be a permutation in S_n expressed in d.c. form, and let τ be an arbitrary element of S_n. Then the conjugate $\tau\sigma\tau^{-1}$ of σ is

$$(\tau(a_1)\ \tau(a_2)\ldots\tau(a_r))(\tau(b_1)\ldots\tau(b_s))(\tau(c_1)\ldots\tau(c_t))\ldots$$

(obtainable by going through the given expression for σ, replacing each object where it occurs by its image under τ).

The proof is simply a matter of verifying that $\tau\sigma\tau^{-1}$ has the alleged effect on the objects. E.g. it *is* true that $\tau\sigma\tau^{-1}$ maps $\tau(a_1)$ to $\tau(a_2)$, because

$$\begin{aligned}
(\tau\sigma\tau^{-1})(\tau(a_1)) &= \tau[\sigma(\tau^{-1}(\tau(a_1)))] \\
&= \tau[\sigma(a_1)] \quad (\tau^{-1}(\tau(x)) \text{ being always } x) \\
&= \tau(a_2);
\end{aligned}$$

and similarly $\tau\sigma\tau^{-1}$ maps $\tau(a_2)$ to $\tau(a_3)$, etc., etc.

For a numerical illustration of 57.1, take

$$\sigma = (1\ 4\ 7)(2\ 6\ 3)(5), \quad \tau = (1\ 6)(2\ 7\ 5\ 4)(3) \quad (\text{in } S_7).$$

To obtain $\tau\sigma\tau^{-1}$, we simply go through the given expression for σ as indicated above: the 1 gets replaced by $\tau(1)$, i.e. 6; the 4 gets replaced by $\tau(4)$, i.e. 2; etc. Once one is familiar with the procedure, one can immediately write down

$$\tau\sigma\tau^{-1} = (6\,2\,5)(7\,1\,3)(4),$$

or (same thing in a less arbitrary order)

$$\tau\sigma\tau^{-1} = (1\,3\,7)(2\,5\,6)(4).$$

A significant corollary of 57.1 is:

57.2 In S_n two permutations are conjugate to each other iff they have the same cycle pattern. Consequently, each conjugacy class in S_n consists of all the permutations with a particular cycle pattern.

Proof. It is clear from 57.1 (and illustrated in the above numerical example) that if two permutations are conjugate to each other in S_n, then they have the same cycle pattern. It remains to demonstrate the truth of the converse—i.e. the fact that permutations with the same cycle pattern are conjugate to each other.

For this purpose consider two permutations in S_n with the same cycle pattern—say

$$\sigma = (a_1 a_2 \ldots a_r)(b_1 b_2 \ldots b_s) \ldots \underbrace{(f_1)(f_2) \ldots (f_t)}_{[\text{fixed objects}]}$$

and

$$\sigma^* = (a_1^* a_2^* \ldots a_r^*)(b_1^* b_2^* \ldots b_s^*) \ldots (f_1^*)(f_2^*) \ldots (f_t^*).$$

(That it is possible to write them out in this way—with the first cycles in the two products of common length, the second cycles of common length, etc.—is an obvious consequence of their having the same cycle pattern.) Let τ be the element of S_n defined by

$$\tau(a_1) = a_1^*, \tau(a_2) = a_2^*, \ldots, \tau(a_r) = a_r^*, \tau(b_1) = b_1^*, \ldots, \tau(f_t) = f_t^*.$$

(This defines τ satisfactorily since $a_1, a_2, \ldots, f_t$ are the objects $1, 2, \ldots, n$ in some order, as are $a_1^*, a_2^*, \ldots, f_t^*$.) By 57.1, $\tau\sigma\tau^{-1} = \sigma^*$, and therefore σ and σ^* are conjugate in S_n.

This completes proof of the result.

When we with to analyse how (for a particular value of n) the group S_n partitions into conjugacy classes, it is obviously (in view of 57.2) very useful to be able to say how many elements there are with any specified cycle pattern. The answer is in every case obtainable by means of the following general formula.

57.3 The number of elements in S_n with cycle pattern $1^{\lambda_1} . 2^{\lambda_2} . 3^{\lambda_3} \ldots$ is

$$\frac{n!}{1^{\lambda_1} 2^{\lambda_2} 3^{\lambda_3} \ldots \lambda_1! \, \lambda_2! \, \lambda_3! \ldots}.$$

Outline of proof. A permutation of cycle pattern $1^{\lambda_1} . 2^{\lambda_2} . 3^{\lambda_3} \ldots$ is produced by filling the n objects into the blanks in the pattern

$$\underbrace{(\cdot)(\cdot)\ldots(\cdot)}_{\lambda_1} \; \underbrace{(\cdot\cdot)(\cdot\cdot)\ldots(\cdot\cdot)}_{\lambda_2} \; \underbrace{(\cdot\cdot\cdot)(\cdot\cdot\cdot)\ldots(\cdot\cdot\cdot)}_{\lambda_3} \ldots .$$

One could say that the number of ways of filling these n blanks with the n objects is $n!$, but then one must take account of the fact that for any one permutation σ, there are several ways of filling the blanks to produce σ. This is because of the possibility of varying

(a) the starting-point in each cycle that appears in σ, and

(b) the order of the λ_j cycles of length j (for each j)

when one writes down σ in d.c. form. The number of permutations of cycle pattern $1^{\lambda_1} . 2^{\lambda_2} . 3^{\lambda_3} \ldots$ is therefore $n!/k$, where k is the number of different ways that there are (on account of the variability of (a) and (b)) of writing in d.c. form a permutation σ of that cycle pattern.

The variability of (a) allows each cycle of length r to be written in r different ways and therefore allows an l-fold variability in the d.c. form expression for σ, where

$$l = \underbrace{1 \times 1 \times \ldots \times 1}_{\lambda_1 \text{ factors}} \times \underbrace{2 \times 2 \times \ldots \times 2}_{\lambda_2 \text{ factors}} \times \underbrace{3 \times 3 \times \ldots \times 3}_{\lambda_3 \text{ factors}} \times \ldots = 1^{\lambda_1} 2^{\lambda_2} 3^{\lambda_3} \ldots .$$

The variability of (b) means that, for each j, the λ_j cycles of length j can be arranged amongst themselves in $\lambda_j!$ ways and, therefore, that there is altogether an m-fold variability in the d.c. form expression for σ, where $m = \lambda_1! \lambda_2! \lambda_3! \ldots$.

Hence the total number, k, of ways of writing σ in d.c. form is given by

$$k = l \times m = 1^{\lambda_1} 2^{\lambda_2} 3^{\lambda_3} \ldots \lambda_1! \lambda_2! \lambda_3! \ldots;$$

and hence the number of permutations in S_n with cycle pattern $1^{\lambda_1} . 2^{\lambda_2} . 3^{\lambda_3} \ldots$ $(n!/k)$ is as claimed.

As a numerical example of the use of 57.3, let us calculate the number of permutations of cycle pattern $1^2 . 3^2 . 4$ in S_{12}. By 57.3, the number is

$$\frac{12!}{1^2 \times 3^2 \times 4 \times 2! \times 2!},$$

which works out as 3 326 400.

A use of 57.3 of less transient interest is the obtaining of details of how, for a particular value of n, S_n partitions into conjugacy classes. Let us illustrate this first with the case of S_4, which is the subject of the following table. There is one row for each conjugacy class of S_4. Because of 57.2, we can specify each conjugacy class by the common cycle pattern of its elements: thus, for example, in the table the "class $1^2 . 2$" means the conjugacy class consisting of all the permutations with cycle pattern $1^2 . 2$. It is the third column of the table (the orders of the conjugacy classes) that 57.3 helps us to fill in. The final column, giving the common period of the elements of each class, is easy to complete in the light of 56.1.

Class	Example element	Order of class	Period of elements
1^4	ι	1	1
$1^2 . 2$	$(2\,3)$	6	2
2^2	$(1\,3)(2\,4)$	3	2
$1 . 3$	$(1\,4\,2)$	8	3
4	$(1\,2\,4\,3)$	6	4

(Total 24)

The corresponding table for S_5 is:

Class	Example element	Order of class	Period of elements
1^5	ι	1	1
$1^3 . 2$	$(1\,5)$	10	2
$1 . 2^2$	$(1\,5)(2\,3)$	15	2
$1^2 . 3$	$(1\,4\,5)$	20	3
$2 . 3$	$(1\,4)(2\,5\,3)$	20	6
$1 . 4$	$(2\,4\,5\,3)$	30	4
5	$(1\,3\,4\,5\,2)$	24	5

(Total 120)

In the case of S_4, an interesting easily verified fact is that the set V consisting of the identity and the 3 elements of cycle pattern 2^2 is a subgroup. This subgroup is a four-group, since all its non-identity elements have period 2. Moreover, since V is the union of two of the conjugacy classes of S_4, it is clearly the case that

$$\sigma \in V \text{ and } \tau \in S_4 \;\Rightarrow \tau\sigma\tau^{-1} \in V;$$

i.e. (cf. 49.3) $V \lhd S_4$. This is noteworthy because it turns out that non-trivial proper normal subgroups of symmetric groups are rarities, and the reader may be intrigued to be told (though no attempt will be made to prove it here) that the subgroup V of S_4 is the only instance of a non-trivial normal subgroup of a symmetric group with index greater than 2.

58. Arrangements of the objects $1, 2, \ldots, n$

By an **arrangement** of the objects $1, 2, \ldots, n$, we simply mean a sequence consisting of these objects written down in some particular order. E.g., in the case $n = 7$,

$$3521764 \tag{*}$$

is an arrangement of the objects.

The considerations of arrangements in this section may appear to constitute a digression from the theory of symmetric groups and permutations. However, there is an obvious connection, which will be exploited in §59: for each permutation $\sigma \in S_n$ naturally gives rise to a certain arrangement of $1, 2, \ldots, n$, namely the arrangement

$$\sigma(1)\sigma(2)\sigma(3)\ldots\sigma(n),$$

which we shall term the *arrangement associated with* σ and denote by A_σ. Observe that it is the sequence of numbers in the second row of the $2 \times n$ matrix notation for σ.

Returning to the above example arrangement (*), consider how the objects 4 and 7 appear in it. Specifically, notice that 7 appears before 4, whereas in the natural ordering of the positive integers (i.e. $1, 2, 3, 4, 5, 6, 7, \ldots$) 4 comes before 7. We describe this situation by saying that the pair $\{4, 7\}$ is *inverted* in the arrangement (*). It is an easy matter to list all the pairs which are inverted in (*). These pairs are

$$\{1, 2\}, \{1, 3\}, \{1, 5\}, \{2, 3\}, \{2, 5\}, \{4, 5\}, \{4, 6\}, \{4, 7\}, \{6, 7\}$$

making 9 inverted pairs altogether.

Obviously such a count of inverted pairs could be made for any given arrangement of the objects $1, 2, \ldots, n$ (whatever the value of n), and this brings us to the main definition of this section.

For an arbitrary arrangement A of the objects, we define the ε-**value** of A (denoted by $\varepsilon(A)$) to be $(-1)^m$, where m is the total number of inverted pairs in the arrangement A.

So, for example, since (as noted above) there are 9 inverted pairs in the arrangement 3521764 of $1, 2, \ldots, 7$,

$$\varepsilon(3521764) = (-1)^9 = -1.$$

Clearly the value of $\varepsilon(A)$ is in every case 1 or -1, and its value is an indication of whether the number of inverted pairs in A is even or odd. Therefore, if a given arrangement were altered so as to change the number of inverted pairs by precisely 1, the ε-value would be reversed—i.e. changed from $+1$ to -1, or vice versa.

At a first reading it might be thought that the ε-value of an arrangement is a somewhat arcane thing to consider. However, as will be seen before the end of the chapter, it starts us out on a trail which leads to interesting discoveries concerning symmetric groups and their elements.

We conclude the present section with a result which discloses what in retrospect may well be adjudged to be the key fact about ε-values of arrangements.

58.1 Suppose that A is a given arrangement of the objects $1, 2, \ldots, n$, and that arrangement A' is obtained from A by interchanging two objects—say a and b. Then $\varepsilon(A') = -\varepsilon(A)$.

(We can usefully re-express this by saying that the effect of interchanging two objects in an arrangement is to reverse the ε-value.)

Proof. It is fairly easy to see that the assertion is true when the interchanged objects a and b are adjacent in the arrangement A. In this case, when a and b are interchanged, the pair $\{a, b\}$ is the only pair of objects whose order of appearance is disturbed. Hence (depending on whether or not the pair $\{a, b\}$ is inverted in A) the effect of interchanging a and b is either to decrease or to increase the number of inverted pairs by 1. So in any event the number of inverted pairs is changed by precisely 1, and therefore the ε-value is reversed.

We now move on to the less easy case where there are k ($\geqslant 1$) objects $x_1, x_2, \ldots, x_k$ between a and b in the arrangement A. Thus A is of the form

$$\ldots a \, x_1 \, x_2 \ldots x_k \, b \ldots .$$

We observe that the interchange of a and b can be brought about by a sequence of adjacent interchanges (i.e. interchanges of adjacent objects), which we now describe and which the diagram below should make clear. First interchange a and x_1 (i.e., in effect, pass a over x_1); then interchange a and x_2 (i.e. pass a over x_2); and proceed in this way until a has been passed over all the x_js and finally over b. Now interchange b and x_k (i.e. pass b leftwards over x_k); then pass b over x_{k-1}; and continue in this way until b has been passed leftwards over all the x_js (ending with x_1)—at which point b occupies the position a had at first, and vice versa.

$$\ldots a \quad x_1 \quad x_2 \quad \ldots \quad x_{k-1} \quad x_k \quad b \quad \ldots$$

$\Big\downarrow$ $(k+1)$ adjacent interchanges

$$\ldots \quad x_1 \quad x_2 \quad \ldots \quad x_{k-1} \quad x_k \quad b \quad a \quad \ldots$$

$\Big\downarrow$ k adjacent interchanges

$$\ldots \quad b \quad x_1 \quad x_2 \quad \ldots \quad x_{k-1} \quad x_k \quad a \quad \ldots$$

As indicated in the diagram, the first stage (where a is passed rightwards over all the x_js and then b) involves $(k+1)$ adjacent interchanges, and the second stage (where b is passed leftwards over all the x_js) involves k adjacent interchanges. The whole process, therefore, involves altogether $(2k+1)$ adjacent interchanges. Thus the interchange of a and b has been brought about by an odd number of adjacent interchanges, each of which, we know from the first paragraph of the proof, reverses the ε-value. The net result of the odd number of reversals is a reversal of the ε-value.

The stated result is now proved in all cases.

59. The alternating character, and alternating groups

The alternating character on the group S_n is the mapping χ from S_n to the multiplicative group $\{1, -1\}$ defined by

$$\chi(\sigma) = \varepsilon(A_\sigma) \qquad (\sigma \in S_n),$$

where A_σ is the arrangement of $1, 2, \ldots, n$ associated with σ, i.e. the arrangement $\sigma(1)\,\sigma(2)\ldots\sigma(n)$.

A permutation in S_n is described as an **odd** permutation if $\chi(\sigma) = -1$ and as an **even** permutation if $\chi(\sigma) = 1$.

For a numerical illustration of these ideas, suppose that σ is the permutation

$$\begin{pmatrix} 1 & 2 & 3 & 4 & 5 & 6 & 7 \\ 3 & 5 & 2 & 1 & 7 & 6 & 4 \end{pmatrix}$$

in S_7. Then

$$\begin{aligned} \chi(\sigma) &= \varepsilon(A_\sigma) \quad \text{(by the above definition of } \chi\text{)} \\ &= \varepsilon(3521764) \\ &= -1 \quad \text{(as seen in a numerical illustration in §58),} \end{aligned}$$

and hence σ is an odd permutation.

The value of $\chi(\sigma)$ is easily seen in the case $\sigma = \iota$, the identity element of S_n:

59.1 $\chi(\iota) = 1$.

For: the arrangement associated with ι is $1234\ldots n$, in which, obviously, there are no inverted pairs, and hence

$$\chi(\iota) = \varepsilon(1234\ldots n) = (-1)^0 = 1.$$

We shall shortly be in a position to prove the important fact that the alternating character mapping χ is a homomorphism (from the group S_n to the group $\{1, -1\}$). Before that come some preliminary results, the most powerful of which is the next proposition. It is the immediate outcome of applying the major result on arrangements (58.1) to permutations in the obvious way.

59.2 Let $\sigma \in S_n$, and let τ be a transposition in S_n. Then

$$\chi(\tau\sigma) = -\chi(\sigma).$$

Proof. Since σ is a bijection, the objects interchanged by the transposition τ are equal to $\sigma(a)$ and $\sigma(b)$ for certain objects a and b, with, suppose, $a < b$. Moreover, since τ moves only $\sigma(a)$ and $\sigma(b)$, it is clear that $\tau\sigma(c) = \sigma(c)$ for every object c other than a and b. Therefore, the arrangements A_σ and $A_{\tau\sigma}$ are respectively as follows:

$$\sigma(1)\,\sigma(2)\,\sigma(3)\ldots\sigma(a)\ldots\sigma(b)\ldots\sigma(n)$$
$$\sigma(1)\,\sigma(2)\,\sigma(3)\ldots\sigma(b)\ldots\sigma(a)\ldots\sigma(n),$$

one obtainable from the other by interchange of $\sigma(a)$ and $\sigma(b)$. By 58.1, $\varepsilon(A_{\tau\sigma}) = -\varepsilon(A_\sigma)$: i.e. $\chi(\tau\sigma) = -\chi(\sigma)$, as asserted.

From 59.1 and 59.2, it follows that, if τ is any transposition, then

$$\chi(\tau) = \chi(\tau\iota) = -\chi(\iota) = -1.$$

Thus:

59.3 Every transposition is an odd permutation.

A more general result is:

59.4 If $\sigma \in S_n$ and σ is expressible as the product of l transpositions, then $\chi(\sigma) = (-1)^l$.

Proof. Suppose that $\sigma = \tau_1\tau_2\ldots\tau_l$, where $\tau_1, \tau_2, \ldots, \tau_l$ are transpositions in S_n. Then

$$\begin{aligned}
\chi(\sigma) &= \chi(\tau_1\tau_2\tau_3\ldots\tau_{l-1}\tau_l)\\
&= (-1)\times\chi(\tau_2\tau_3\ldots\tau_{l-1}\tau_l) \quad \text{(by 59.2)}\\
&= (-1)^2\times\chi(\tau_3\tau_4\ldots\tau_{l-1}\tau_l) \quad \text{(by 59.2 again)},\\
&= \ldots\\
&= (-1)^{l-1}\times\chi(\tau_l) \quad \text{(after } l-1 \text{ such steps)}\\
&= (-1)^l \quad \text{(by 59.3)}.
\end{aligned}$$

This proves the result.

We now come to the earlier announced crucial fact about the alternating character mapping.

59.5 The alternating character $\chi: S_n \to \{1, -1\}$ is a homomorphism.

Proof. Let ρ, σ be arbitrary elements of S_n.

By 54.2, each of ρ and σ can be expressed as a product of transpositions:

say $\rho = \tau_1 \tau_2 \ldots \tau_l$ and $\sigma = \tau_{l+1} \tau_{l+2} \ldots \tau_{l+m}$, where $\tau_1, \tau_2, \ldots, \tau_{l+m}$ are transpositions. Our notation gives us ρ and σ as the products of l transpositions and of m transpositions, respectively; and hence, by 59.4, $\chi(\rho) = (-1)^l$, and $\chi(\sigma) = (-1)^m$. Further, $\rho\sigma = \tau_1 \tau_2 \ldots \tau_{l+m}$, the product of $(l+m)$ transpositions, so that we have

$$\chi(\rho\sigma) = (-1)^{l+m} = (-1)^l \times (-1)^m = \chi(\rho)\chi(\sigma).$$

Since ρ, σ were arbitrary in S_n, this proves that χ is a homomorphism.

The kernel, ker χ, of the homomorphism χ is by definition

$$\{\sigma \in S_n : \chi(\sigma) = 1\}.$$

So ker χ consists precisely of the even permutations in S_n. It follows at once (by 49.4) that the even permutations in S_n form a normal subgroup of S_n. This normal subgroup comprising all the even permutations in S_n is called the **alternating group** of degree n, and we shall denote it by A_n.

To ascertain the order of A_n, we prove the following general proposition about the possible proportions of even permutations in an arbitrary subgroup of S_n.

59.6 Suppose that G is a subgroup of S_n ($n \in \mathbb{N}$). Then either G consists entirely of even permutations (i.e. $G \subseteq A_n$), or else the even permutations in G form a subgroup of index 2 in G.

Proof. Let K be the set of even permutations in G. It may be the case that K is the whole of G. Consider henceforth the remaining case where G contains at least one odd permutation, and let ρ denote an odd permutation in G. It suffices to prove that in this case K is a subgroup of G with $|G:K| = 2$.

Let $\chi^* = \chi|_G$, the restriction of χ to G as domain. (See the last paragraph of §23.) Clearly χ^* is a homomorphism, and its kernel is K. So K is a (normal) subgroup of G. Further, by the first isomorphism theorem (52.1),

$$G/K \cong \text{im } \chi^*.$$

It follows that $|G/K| = |\text{im } \chi^*|$, i.e. $|G:K| = |\text{im } \chi^*|$. But, since ι and ρ belong to G and their images under χ^* are 1 and -1, respectively, it is evident that im χ^* is the whole of $\{1, -1\}$. Thus $|\text{im } \chi^*| = 2$, and therefore $|G:K| = 2$. The stated result follows.

A corollary is:

59.7 For $n \geqslant 2$, $|S_n : A_n| = 2$, and so $|A_n| = \frac{1}{2}(n!)$.

Proof. Suppose $n \geqslant 2$. Then (cf. 59.3) S_n does contain at least one odd permutation—e.g. the transposition (1 2). So it follows from 59.6 that $|S_n : A_n| = 2$, as asserted.

Remarks

(a) It is necessary to exclude the case $n = 1$ in the statement of 59.7. Of course, the symmetric group S_1 of degree 1 is a trivial group (of order 1), and so there is seldom if ever any point in considering S_1 in permutation group theory. However, for the record, the one element, ι, of S_1 is an even permutation, and thus the alternating group A_1 is the whole of S_1.

(b) The alternating groups A_2 and A_3 are also of limited interest. On calculating their orders, one sees that they are trivial and cyclic of order 3, respectively (cf. 44.2).

(c) The group of rotational symmetries of a regular icosahedron can be shown to be isomorphic to A_5 which, for this reason, is often called the *icosahedral group*.

(d) The perception of the distinction between odd and even permutations and the discovering of the existence and order of the subgroup A_n of S_n certainly constitute significant advances so far as the theory of permutation groups and its applications are concerned. But in addition there is, as will be indicated in §60, a reason (quite apart from permutation groups specifically) for regarding the groups A_n for $n \geqslant 5$ as having a special status in the general theory of the structure of finite groups.

The final objective of this section is to show how one can tell from its cycle pattern whether a given permutation is even or odd. As will be demonstrated, this is possible as soon as one is aware of the following fact about cycles.

59.8 A cycle of even length is an odd permutation, and a cycle of odd length is an even permutation.

Proof. Consider an arbitrary cycle $\alpha = (a_1 a_2 \ldots a_r)$ of length r ($\geqslant 2$) in S_n. The key to the proof is the observation (easily verified once it comes to one's attention) that α is equal to

$$(a_1 a_r)(a_1 a_{r-1}) \ldots (a_1 a_4)(a_1 a_3)(a_1 a_2).$$

a product of $(r-1)$ transpositions. Hence, by 59.4,

$$\chi(\alpha) = (-1)^{r-1},$$

and it can be seen that the stated result follows from this.

It is because χ is a homomorphism that 59.8 enables us to determine the evenness or oddness of any permutation given in d.c. form. A specific example will make this clear: let us illustrate with the permutation

$$\sigma = (1\ 3)(2\ 14)(4\ 13\ 8)(5\ 11\ 10\ 6\ 9\ 12) \qquad (\in S_{14}).$$

(σ fixes 7, but in this context we can ignore fixed objects.) By 59.8, the χ-values of the cycles (1 3), (2 14), (4 13 8) and (5 11 10 6 9 12) whose product is σ are respectively

$$-1, -1, 1, \text{ and } -1.$$

Hence, since χ is a homomorphism,

$$\chi(\sigma) = (-1) \times (-1) \times 1 \times (-1) = -1;$$

i.e. σ is an odd permutation.

A little thought will soon make it clear that the general result obtainable through the approach just illustrated is:

59.9 A permutation in S_n is odd iff there is altogether an odd number of cycles of even length in its d.c. form.

Looking back at the tables of conjugacy classes of S_4 and S_5 towards the end of §57, one can now easily see which members of S_4 belong to A_4 and which members of S_5 belong to A_5. From the S_4 table one sees that the group A_4 (which has order 12) consists of the identity element, the 3 permutations of cycle pattern 2^2, and the 8 elements of cycle pattern 1.3. The corresponding story for A_5 (which we shall refer back to in §60) is:

59.10 The group A_5 (which has order 60) consists of the following elements of S_5: the identity, the 15 elements of cycle pattern 1.2^2, the 20 elements of cycle pattern $1^2.3$, and the 24 elements of cycle pattern 5.

60. The simplicity of A_5

A **simple group** means a non-trivial group which has no non-trivial proper normal subgroup.

To see the point of this definition, consider a finite group G which *does* have a non-trivial proper normal subgroup K. Because $K \lhd G$, the group G/K exists; and both of the groups K and G/K are smaller than G. It must be emphasised that it is *not* true that the structures of the two groups K and G/K determine the structure of G. Nevertheless, knowledge of the structures of the groups K and G/K will in general help us towards some purposeful statements about the group G in terms of two smaller groups. Clearly, however, this approach is not open to us in the case of a simple group, and so we may liken simple groups to atoms or fundamental building-bricks in terms of which the structures of other groups may be, to some extent, described. There is, therefore, great interest in simple groups in group theory.

Since every subgroup of an abelian group is normal in that group, a simple abelian group is a non-trivial group with no non-trivial proper subgroups whatsoever. It is not difficult to show (cf. exercise 23 on page 88) that the groups with this property are precisely the cyclic groups of prime order. Thus:

60.1 The abelian simple groups are the cyclic groups of prime order.

The search for non-abelian simple groups is an infinitely harder task. It is a task to which a formidable effort has been devoted by group-theorists in recent decades—very productively, let it be added. Here we shall prove just one modest, but very interesting, result on this theme—that A_5 (which has order 60) is one example of a non-abelian simple group. (It is certainly non-abelian: for example, the cycles (1 2 3) and (1 2 4), which by 59.8 belong to A_5, do not commute with each other.)

As a preliminary we shall investigate in general how an alternating group partitions into conjugacy classes. For this purpose it is necessary to distinguish carefully (for arbitrary $\sigma \in A_n$) between

$$C_{A_n}(\sigma)\text{—the centralizer of }\sigma\text{ in }A_n\text{, i.e. }\{\rho \in A_n : \rho\sigma = \sigma\rho\},$$

and

$$C_{S_n}(\sigma)\text{—the centralizer of }\sigma\text{ in }S_n\text{, i.e. }\{\rho \in S_n : \rho\sigma = \sigma\rho\},$$

and between

$$\mathrm{cl}_{A_n}(\sigma)\text{—the conjugacy class of }\sigma\text{ in }A_n\text{, i.e. }\{\rho\sigma\rho^{-1} : \rho \in A_n\},$$

and

$$\mathrm{cl}_{S_n}(\sigma)\text{—the conjugacy class of }\sigma\text{ in }S_n\text{, i.e. }\{\rho\sigma\rho^{-1} : \rho \in S_n\}.$$

We know (cf. 57.2) that $\mathrm{cl}_{S_n}(\sigma)$ consists precisely of the permutations in S_n with the same cycle pattern as σ, and it is readily seen that:

60.2 $\mathrm{cl}_{A_n}(\sigma) \subseteq \mathrm{cl}_{S_n}(\sigma)$ $(\sigma \in A_n)$,

since $\{\rho\sigma\rho^{-1} : \rho \in A_n\} \subseteq \{\rho\sigma\rho^{-1} : \rho \in S_n\}$, obviously, as A_n is a subset of S_n. However, it does happen in certain instances that $\mathrm{cl}_{A_n}(\sigma)$ is a proper subset of $\mathrm{cl}_{S_n}(\sigma)$, and the next result gives details of this. Attention here is restricted to $n \geqslant 2$, since the proof to be given does not cover the case $n = 1$ and since that case is of zero interest.

60.3 Let $\sigma \in A_n (n \geqslant 2)$. If there is an odd permutation in S_n that commutes with σ, then $\mathrm{cl}_{A_n}(\sigma) = \mathrm{cl}_{S_n}(\sigma)$. Otherwise, $\mathrm{cl}_{A_n}(\sigma)$ consists of precisely half the elements of $\mathrm{cl}_{S_n}(\sigma)$.

Proof. We begin by remarking that we must have one or other of the following two cases.

Case 1: there is an odd permutation in S_n that commutes with σ, i.e. there is an odd permutation in $C_{S_n}(\sigma)$. In this event (by 59.6) precisely half the elements of $C_{S_n}(\sigma)$ are even permutations.

Case 2: there is no odd permutation in S_n that commutes with σ, i.e. all the elements of $C_{S_n}(\sigma)$ are even permutations.

Since, clearly, $C_{A_n}(\sigma)$ consists of the even permutations in S_n that commute with σ, it is apparent that

$$|C_{A_n}(\sigma)| = \begin{cases} \frac{1}{2}|C_{S_n}(\sigma)| & \text{in case 1,} \\ |C_{S_n}(\sigma)| & \text{in case 2.} \end{cases}$$

Now, in any event,

$$|\mathrm{cl}_{A_n}(\sigma)| = \frac{|A_n|}{|C_{A_n}(\sigma)|} \qquad \text{(by 48.2)}$$

$$= \frac{\frac{1}{2}|S_n|}{|C_{A_n}(\sigma)|} \qquad \text{(cf. 59.7).}$$

So, in case 1,

$$|\mathrm{cl}_{A_n}(\sigma)| = \frac{\frac{1}{2}|S_n|}{\frac{1}{2}|C_{S_n}(\sigma)|} = \frac{|S_n|}{|C_{S_n}(\sigma)|} = |S_n : C_{S_n}(\sigma)| = |\mathrm{cl}_{S_n}(\sigma)|,$$

while, in case 2,

$$|\mathrm{cl}_{A_n}(\sigma)| = \frac{\frac{1}{2}|S_n|}{|C_{S_n}(\sigma)|} = \frac{1}{2}|S_n : C_{S_n}(\sigma)| = \frac{1}{2}|\mathrm{cl}_{S_n}(\sigma)|;$$

and, in view of 60.2, these conclusions justify what is stated in the present proposition.

Another way of expressing what 60.3 tells us is this: if X is a conjugacy class in S_n whose members are even permutations, then either (1) X is also a conjugacy class in A_n—which happens if the centralizer of an element of X contains an odd permutation, or (2) X splits into two conjugacy classes of A_n, each of order $\frac{1}{2}|X|$—which happens if the centralizer of each element of X consists entirely of even permutations.

Let us now apply this knowledge to A_5, whose membership was listed in terms of S_5-conjugacy classes in 59.10. Obviously, an S_5-conjugacy class of odd order cannot split into two equal-sized subsets; and so it is immediately apparent that the 15 elements of cycle pattern $1 . 2^2$ must form a single conjugacy class in A_5, as of course does the identity element. On the other hand, it is obvious (since the order of a conjugacy class in a group G must divide $|G|$) that the 24 elements of cycle pattern 5 cannot form a single conjugacy class in A_5 (which has order 60); and so these elements must form two A_5-conjugacy classes, each of order 12. That leaves the set of 20 elements of cycle pattern $1^2 . 3$. One such element is the cycle $\alpha = (1\,2\,3)$; and it is not difficult to spot that $(4\,5)$ is an odd permutation commuting with α. So, in fact, by 60.3 the 20 elements of cycle pattern $1^2 . 3$ form a single conjugacy class in A_5.

To sum up, the following conclusion has been reached.

60.4 The group A_5 partitions into 5 conjugacy classes, namely:

(1) $\{\iota\}$,
(2) the set of 15 elements of cycle pattern 1.2^2,
(3) the set of 20 elements of cycle pattern $1^2.3$,
(4) a set comprising 12 of the elements of cycle pattern 5, and
(5) the set consisting of the other 12 elements of cycle pattern 5.

The result which is the climax of §60 can now be proved, namely:

60.5 The group A_5 is simple.

Proof. Suppose that A_5 is not simple. Then A_5 contains a normal subgroup K whose order lies strictly between 1 and 60 and is (by Lagrange's theorem) a divisor of 60.

Because $K \lhd A_5$,

$$\sigma \in K \Rightarrow \tau\sigma\tau^{-1} \in K \text{ for all } \tau \in A_5 \qquad \text{(cf. 49.3),}$$

i.e. $\sigma \in K \Rightarrow$ every element of $\mathrm{cl}_{A_5}(\sigma)$ lies in K.

Hence it is apparent that K must be the union of some entire conjugacy classes of A_5; and of course one of these is $\{\iota\}$, since the subgroup K must contain the identity element. Different conjugacy classes being disjoint, it follows that $|K|$ is equal to the total of the orders of certain conjugacy classes of A_5 including $\{\iota\}$. However, from 60.4 one can soon check that no such total is equal to a divisor of 60 strictly between 1 and 60, and thus we have a contradiction of what was said earlier about $|K|$. From this contradiction the stated result follows.

Two supplementary facts are worth mentioning which help to put 60.5 in perspective. The first is that A_5 is the *smallest* of the non-abelian simple groups. (The proof that there is no non-abelian simple group with order $\leqslant 59$ could be set as a not too difficult exercise at a slightly more advanced stage in the study of group theory than is reached here.) The second supplementary fact is a generalization of 60.5—that A_n is simple for *every* $n \geqslant 5$. Because of the significance of simple groups (as outlined at the beginning of §60), it follows that alternating groups are of interest in group theory well beyond that part of the subject which deals purely with permutations. (There exists an elementary, albeit somewhat long, proof of the simplicity of A_n for all $n \geqslant 5$. A version of this elementary proof may be found in [3].)

At this point the section of the book devoted to groups draws to a close, since to go further would be at variance with the book's aim to provide an introduction, of reasonable length, to abstract algebra as a whole. The last four chapters have introduced the student to the basic concepts of group theory such as period of an element, cyclic groups, subgroups, cosets, homomorphisms and their kernels, quotient groups, and matters of fundamental importance in permutation group theory. An understanding of these things equips the student to advance to higher levels in the study of groups if he wishes. It will also help him in the study of other species of algebraic systems, since concepts closely analogous to those of elementary group theory often loom large in the study of other systems—a point which will emerge clearly in the discussion of rings in the next chapter.

The further theory of groups is a particularly attractive part of mathematics. In it one comes across theorems of great power and elegance on group structure (e.g. theorems about the existence of and relationships between subgroups of a group—both general theorems and theorems about special classes of groups that one can identify as in some sense nice). However, many significant problems about groups are very formidable problems, and it should not be imagined that complete solutions to all these problems are known to mankind.

Students keen to read more about group theory will find plenty to interest them in textbooks on the subject such as those by I. D. Macdonald, W. R. Scott, and J. Rose ([4], [5], and [6], respectively).

EXERCISES ON CHAPTER NINE

1. In S_6, let

$$\sigma = \begin{pmatrix} 1 & 2 & 3 & 4 & 5 & 6 \\ 5 & 4 & 1 & 3 & 6 & 2 \end{pmatrix}, \qquad \tau = \begin{pmatrix} 1 & 2 & 3 & 4 & 5 & 6 \\ 6 & 5 & 2 & 4 & 3 & 1 \end{pmatrix}.$$

Write down, in the same form, each of the permutations $\sigma\tau, \tau\sigma, \sigma^{-1}, \tau^{-1}, \tau^2$.

2. Express in d.c. form the permutations σ and τ given in exercise 1.

3. Write down the cube and the inverse of the cycle (1 5 3 2 4).

4. In S_{10}, let

$$\sigma = (1\,4)(2\,9\,8\,6)(3\,10\,5\,7), \qquad \tau = (1\,7\,5)(2\,6)(3\,9)(8\,10)(4).$$

(a) Write down the cycle patterns of σ and τ.

(b) Without translating σ or τ into any other form, find in d.c. form each of the permutations $\sigma\tau, \tau\sigma, \sigma^{-1}, \tau^{-1}, \sigma^2, \sigma^3, (\sigma\tau)^3$.

(c) What are the periods of σ, τ, and $\sigma\tau$?

5. In S_8, let

$$\sigma = \begin{pmatrix} 1 & 2 & 3 & 4 & 5 & 6 & 7 & 8 \\ 3 & 8 & 4 & 6 & 5 & 1 & 2 & 7 \end{pmatrix}, \qquad \tau = \begin{pmatrix} 1 & 2 & 3 & 4 & 5 & 6 & 7 & 8 \\ 7 & 5 & 3 & 1 & 8 & 6 & 4 & 2 \end{pmatrix}.$$

(a) Express each of the permutations $\sigma, \tau, \sigma\tau, \tau\sigma\tau^{-1}$ in d.c. form. (Be sure to use 57.1 in the last case.)

(b) State the periods of σ and τ.

(c) Find the number of conjugates of τ in S_8.

(d) Find a permutation $\gamma \in S_8$ such that $\gamma^2 = \tau$ and γ moves all the objects $1, 2, 3, \ldots, 8$.

6. In S_7, let

$$\sigma = (1\,4)(2\,7\,5\,3)(6), \qquad \tau = (2\,4)(3\,6\,7\,5)(1),$$

so that σ and τ have the same cycle pattern and are therefore conjugate in S_7. Write down a permutation $\gamma \in S_7$ such that $\gamma\sigma\gamma^{-1} = \tau$.

7. What cycle patterns are possible for a permutation in S_7 that fixes none of the objects? Hence find the total number of such permutations in S_7.

8. Find (a) the number of elements of period 3 in S_7, (b) the number of elements of period 6 in S_8.

9. Show that in S_6 the elements of period 3 partition into two equal-sized conjugacy classes. Show also that if σ is any element of period 3 in S_6, $C_{S_6}(\sigma)$ contains an element of period 6.

10. Let α be a cycle in S_n (n arbitrary). Suppose that there is an object a such that $\alpha(a) \neq a$ but $\alpha^r(a) = a$ for a certain positive integer r. Show that $\alpha^r = \iota$.

11. A certain permutation σ in S_9 has cycle pattern $1^2 . 2^2 . 3$, and the permutation γ satisfies $\gamma^2 = \sigma$. How many different cycle patterns might γ have?

12. In S_9, let $\sigma = (1\,3)(2\,9)(4\,7)(5\,8\,6)$. Write down a permutation of cycle pattern 3^3 that commutes with σ. (Note that $\tau\sigma = \sigma\tau$ is equivalent to $\tau\sigma\tau^{-1} = \sigma$.)

13. Let α be a cycle of length n in S_n (n arbitrary). Show, by considering $|C_{S_n}(\alpha)|$, that $C_{S_n}(\alpha) = \langle \alpha \rangle$.

14. By counting inverted pairs, find the ε-value of the arrangement 25341.

15. In S_{10}, let

$$\sigma = \begin{pmatrix} 1 & 2 & 3 & 4 & 5 & 6 & 7 & 8 & 9 & 10 \\ 5 & 6 & 4 & 3 & 10 & 2 & 1 & 8 & 7 & 9 \end{pmatrix}, \quad \tau = \begin{pmatrix} 1 & 2 & 3 & 4 & 5 & 6 & 7 & 8 & 9 & 10 \\ 3 & 4 & 10 & 2 & 7 & 1 & 6 & 9 & 8 & 5 \end{pmatrix}.$$

Express each of σ and τ in d.c. form, and hence determine whether each of σ, τ is an even or odd permutation.

16. If σ, τ are respectively odd and even permutations in S_n, which of the permutations

$$\sigma\tau, \sigma^2, \tau^2, \sigma^3, \sigma^{-1}, \tau^{-1}, \tau\sigma\tau^{-1}$$

belong to A_n?

17. (a) What is the smallest value of n such that there is an element of period 18 in S_n? (b) What is the smallest value of n such that there is an element of period 18 in A_n?

18. Let $\sigma = (1\ 2\ 3)(4\ 5\ 6)$. Find (a) an odd permutation, (b) a non-identity even permutation in $C_{S_6}(\sigma)$.

19. Find the details of how A_6 partitions into conjugacy classes.

20. Verify that if a, b are objects and $1, a, b$ are all different, then

$$(1\ a)(1\ b)(1\ a) = (a\ b).$$

Deduce that (a) for $n \geq 2$, every element of S_n can be expressed in terms of the transpositions of the form $(1\ c)$ $(c \neq 1)$, and (b) for $n \geq 3$, every element of A_n can be expressed as a product of cycles of length 3.

21. This is a nice application of 59.6 to prove that a group of order $2n$, where n is odd, contains a subgroup of order n. Let G be such a group, and consider the isomorphic group $\theta(G)$, where θ is the isomorphism introduced in the proof of Cayley's theorem (47.9). By exercise 13 on chapter 6, G contains an element c of period 2. Show that in $\theta(G)$ the permutation λ_c (which must also have period 2 and is to be regarded as a permutation of the n objects in G) has cycle pattern 2^n. Deduce that $\theta(G)$ contains a subgroup of order n, and therefore so does G.

22. Deduce from exercise 21 that if G is a simple group of even order greater than 2, then $4 \,|\, |G|$. (Note: there is a famous theorem, the Feit–Thomson theorem, proof of which was first published in 1963, which tells us that every non-abelian simple group has even order.)

23. A subgroup G of S_n is described as *transitive* iff for every ordered pair (a, b) of objects there exists $\sigma \in G$ such that $\sigma(a) = b$.
(a) Show that, for every $n \geq 3$, A_n is transitive.
(b) Show that if G is a transitive abelian subgroup of S_n, then each non-identity element of G moves all the objects.

24. Let G be a subgroup of S_n.
(a) Show that the relation $\sim$ defined on the set of objects by

$$a \sim b \Leftrightarrow \exists\, \sigma \in G \text{ s.t. } \sigma(a) = b \qquad (a, b \in \{1, \ldots, n\})$$

is an equivalence relation. The $\sim$-class of object a is termed the *orbit* of a. Notice that the number of objects in the orbit of a is (by exercise 17 on chapter 7) equal to $|G:G_a|$, where G_a is the stabilizer of a in G, i.e. $\{\sigma \in G : \sigma(a) = a\}$.

(*b*) Let a be an object, and let $\sigma \in G$. Show that $G_{\sigma(a)}$, the stabilizer of $\sigma(a)$ in G, is $\sigma G_a \sigma^{-1}$ (G_a being the stabilizer of a).

(*c*) For each $\gamma \in G$, let $F(\gamma)$ denote the number of objects fixed by γ. By counting the number of pairs of the form (γ, a) with a an object and γ a permutation in G fixing a, show that

$$\sum_{\gamma \in G} F(\gamma) = |G| \times k,$$

where k is the number of different orbits under the action of G, i.e. the number of different $\sim$-classes that there are, where $\sim$ is the equivalence relation introduced in part (*a*).

25. (i) Show that, although $6 \,|\, |A_4|$, A_4 contains no subgroup of order 6. Is A_4 a simple group?

(ii) Part (i) reveals that what is sometimes termed "the converse of Lagrange's theorem"—i.e. the statement that if $d \,(\in \mathbb{N})$ divides the order of the finite group G, then G must contain a subgroup of order d—is false. A more sophisticated suggestion along the same lines that might be put forward is: "If $d \,(\in \mathbb{N})$ divides the order of the finite group G and if d and $|G|/d$ are relatively prime, then G contains a subgroup of order d". Prove that this suggestion is also false by making use of the result of exercise 28 on chapter 8 (that every group of order 15 is cyclic).

CHAPTER TEN

RINGS

61. Introduction

In this final chapter, we look at properties of rings, i.e. algebraic systems with two binary operations (addition and multiplication) satisfying certain postulates. For part of the time our concern is with special kinds of rings, e.g. fields, to which the reader was introduced towards the end of chapter 3.

In the early sections of the chapter we discuss features of ring theory which are obviously analogous to features of group theory developed in chapters 6 to 8. For example, we define and discuss subrings, homomorphisms from one ring to another, kernels of homomorphisms. A subring that is the kernel of a homomorphism is what we call an *ideal* (cf. a normal subgroup in group theory) and it turns out that, just as in group theory the set of cosets of a normal subgroup can be turned into a group, so in ring theory the set of cosets of an ideal can be turned into a ring.

Later in the chapter we start exploring the subject of factorization of elements of rings of a certain kind, and thereby begin an attempt to put the factorization properties of the integers into perspective and see how far they can be generalized. Then our final topic is polynomial rings, these being rings that crop up naturally in a wide variety of mathematical situations.

62. The definition of a ring and its elementary consequences

Definition. A non-empty set R is called a **ring** when there are defined on R two binary operations, addition and multiplication, and the following postulates are satisfied:

A: R is an abelian group with respect to addition;

M0: R is closed under multiplication;

M1: multiplication is associative on R;

D: $\forall x, y, z \in R, \begin{cases} x(y+z) = xy + xz \\ \text{and } (y+z)x = yx + zx \end{cases}$ (*distributive* laws)

The postulates A, M0, M1, D are called the **ring axioms.**

Examples of rings include the following systems: $\mathbb{Z}$; $\mathbb{R}$; $\mathbb{C}$; the system of all $n \times n$ matrices with entries in $\mathbb{R}$ (n being a given positive integer); the system $\mathbb{Z}_m$ of congruence classes modulo m (m being a given positive integer, and addition and multiplication being defined as in §18).

Remarks. (*a*) Pedantically speaking, a ring is not just a set R but rather an ordered triple $(R, +, .)$ consisting of a set R and two binary operations defined on that set. Nevertheless (cf. the remark following the definition of a semigroup in §29) it is commonplace to use phraseology such as "the ring R".

(*b*) Clearly, with respect to multiplication, the set of elements in a ring is a semigroup.

(*c*) Ring axiom A tells us that in every ring there is a zero and each element x has a negative, $-x$; and it tells us how zero and negatives behave in addition in a ring. We now show that it is a consequence of the ring axioms that, in multiplication in a ring, zero and negatives behave in a definite (and natural-looking) way.

62.1 In a ring R, $\quad 0 . x = x . 0 = 0$ for every $x \in R$.

Proof. Let x be an arbitrary element of the ring R. We shall prove that $0 . x = 0$. (The proof that $x . 0 = 0$ is similar.)

Since 0 is the identity element in the group $(R, +)$, $0 + 0 = 0$; and so

$$x(0+0) = x . 0$$

i.e. $\qquad\qquad\qquad x . 0 + x . 0 = x . 0 \qquad \text{(by D)}$

$$= x . 0 + 0.$$

By cancellation in the additive group $(R, +)$ it follows that $x . 0 = 0$.

62.2 Let x, y be elements of a ring R. Then

(i) $(-x)y = x(-y) = -(xy)$, $\qquad$ (ii) $(-x)(-y) = xy$.

Proof. (i) $\{x + (-x)\}y = 0 . y$

$$= 0 \qquad \text{(by 62.1)}$$

i.e. by D, $\quad xy + (-x)y = 0$.

By the additive version of 35.3, it follows that $(-x)y = -(xy)$.
It can similarly be proved that $x(-y) = -(xy)$.

(ii) $(-x)(-y) = -\{x(-y)\} \qquad$ (by part (i))

$$= -\{-(xy)\} \qquad \text{(by part (i))}$$

$$= xy \qquad \text{(by the additive version of 35.5)}.$$

63. Special types of ring and ring elements

(1) *Commutative rings*

A **commutative ring** means a ring in which the multiplication is commutative. The most familiar example is $\mathbb{Z}$.

(2) *Rings with unity*

In a ring R, a **unity** (or identity) is a nonzero element 1 with the property that $1x = x1 = x$ for every $x \in R$. Thus a unity in R is an identity, different from zero, in the semigroup $(R, .)$.

By 31.1, it follows that no ring can possess more than one unity.

A ring with a unity is often referred to colloquially as "a ring with a one".

The ring of 2×2 real matrices is an example of a non-commutative ring with a one, the "one" (or unity) being the matrix $I = \begin{bmatrix} 1 & 0 \\ 0 & 1 \end{bmatrix}$.

The set of even integers (with ordinary addition and multiplication) provides an example of a ring without a unity.

(3) *Division rings*

A **division ring** is a (not necessarily commutative) ring with a unity, 1, in which every nonzero element has a multiplicative inverse (i.e. corresponding to each nonzero element x of the ring there is an element x^{-1}, also in the ring, such that $x^{-1}x = xx^{-1} = 1$).

An example of a division ring which is non-commutative is the ring of all complex matrices of the form $\begin{bmatrix} a & b \\ -\bar{b} & \bar{a} \end{bmatrix}$ $(a, b \in \mathbb{C})$. It is easy but tedious to check that the set of all such matrices actually is a non-commutative ring with a one; the point of special interest is that each nonzero element $\begin{bmatrix} a & b \\ -\bar{b} & \bar{a} \end{bmatrix}$ $(a, b \in \mathbb{C}$ and not both zero) of the ring has a multiplicative inverse in the ring, viz. $\begin{bmatrix} c & d \\ -\bar{d} & \bar{c} \end{bmatrix}$, where

$$c = \bar{a}/(|a|^2 + |b|^2), \quad d = -b/(|a|^2 + |b|^2).$$

This division ring, which can be exhibited in a variety of isomorphic versions, is called the *ring of quaternions*.

(4) *Fields*

Arriving on familiar territory by an unfamiliar route, we define a **field** to be a commutative division ring, i.e. a field is a non-empty set F on which are defined two binary operations, viz. addition and multiplication, the following postulates all being satisfied:

A, M0, M1, D as in the definition of a ring (with F replacing R);

M2: multiplication is commutative on F;

M3: there is a unity in F;

M4: each nonzero element of F has a multiplicative inverse in F.

(It is easily seen that this definition of a field coincides with that given in § 19.)

The very well-known algebraic systems $\mathbb{Q}$, $\mathbb{R}$, $\mathbb{C}$ are all fields. And (cf. 19.2) the system $\mathbb{Z}_p$ is a field when p is prime.

(5) Integral domains

Preliminary definition. In a commutative ring R, a nonzero element x is called a **divisor of zero** iff there exists $y \in R - \{0\}$ such that $xy = 0$.

Thus the presence of a divisor of zero in a commutative ring means that the product of two elements of the ring may be zero, even though neither factor is zero: e.g. in the commutative ring $\mathbb{Z}_6$, we have $C_2 C_3 = C_0$ (C_0 being the zero of the ring), while neither C_2 nor C_3 is zero: this shows that C_2 and C_3 are divisors of zero in the ring $\mathbb{Z}_6$.

Main definition. An **integral domain** is a commutative ring with a one in which there are no divisors of zero; i.e. an integral domain is a commutative ring R with a one in which there holds the additional axiom

$$\text{I: For } x, y \in R, \ xy = 0 \Rightarrow x = 0 \text{ or } y = 0 \text{ (or both)}.$$

The ring $\mathbb{Z}$ is a familiar example of an integral domain. Moreover, $\mathbb{Z}$ is an example of an integral domain which is not a field. This remark should be set alongside the fact that:

63.1 Every field is an integral domain.

Proof. Let F be a field. Since F is certainly a commutative ring with a one, the result will follow if we can show that the axiom I holds in the case $R = F$.

Suppose that $xy = 0$ ($x, y \in F$). If $x \neq 0$, x^{-1} exists in F and we have

$$y = 1y = x^{-1}xy = x^{-1}0 = 0 \qquad \text{(using 62.1 at the last step)}.$$

The only remaining case is $x = 0$.

Therefore, for $x, y \in F$, $xy = 0 \Rightarrow x = 0$ or $y = 0$.

Thus I holds for F; and the result follows.

It is helpful to note that, by the law of contraposition, the axiom I is equivalent to

$$x \neq 0 \quad \text{and} \quad y \neq 0 \Rightarrow xy \neq 0 \qquad (x, y \in R)$$

i.e. $\qquad R - \{0\}$ is closed under multiplication.

This version of I makes it easy to see that:

63.2 If R is an integral domain, then $(R - \{0\}, .)$ is a semigroup.

(For, if R is an integral domain, then (1) $R - \{0\} \neq \emptyset$ since $1 \in R - \{0\}$, (2) $R - \{0\}$ is closed under multiplication (cf. above version of I), and (3) multiplication is associative on $R - \{0\}$ since multiplication is associative on R).

An analogous observation about fields, more conveniently made here than in subsection (4), is:

63.3 If F is a field, then $(F - \{0\}, .)$ is a group.

Proof. Let F be a field. We show that the group axioms hold for $F - \{0\}$.

G0: By 63.1 and 63.2, $F - \{0\}$ is closed under multiplication.

G1: Holds, since multiplication is associative on F.

G2: Holds because (since $1 \neq 0$) $1 \in F - \{0\}$.

G3: Corresponding to each $x \in F - \{0\}$, there is (by the definition of a field) an element $x^{-1} \in F$ such that $x^{-1}x = 1$; and in every case $x^{-1} \neq 0$ (since $x^{-1} = 0$ would imply $1 = x^{-1}x = 0x = 0$). Thus each element of $F - \{0\}$ has an inverse in $F - \{0\}$, i.e. G3 holds.

The result follows.

Our final remark on integral domains is that in such systems there holds a cancellation law to the effect that one can cancel a nonzero element from both sides of an equation:

63.4 Suppose that elements a, b, c of an integral domain R satisfy $ab = ac$ and $a \neq 0$. Then $b = c$.

Proof. $a(b - c) = a\{b + (-c)\}$ (by meaning of subtraction in an additive group—see remark (*b*) in § 33)

$$= ab + a(-c) \quad \text{(by ring axiom D)}$$
$$= ab - ac \quad \text{(by 62.2)}$$
$$= 0.$$

Therefore, since R has no divisors of zero, one at least of $a, b - c$ must be zero. However, $a \neq 0$. Hence $b - c = 0$, and hence $b = c$.

(6) *Units in a ring with a one*

Definition. Let R be a ring with a one. A **unit** in R means an element of R with a multiplicative inverse in R (i.e. the element x of R is a unit iff x has an inverse in the semigroup $(R, .)$).

For example, in $\mathbb{Z}$ the units are 1 and -1; in a field every nonzero element is a unit.

A generalization of 63.3 is:

63.5 Let R be a ring with a one, and let U be the set of units in R. Then U is a group with respect to multiplication.

The proof of 63.5 is left as an exercise.

64. Subrings and subfields

Definition 1. Let R be a ring. [This means that the system $(R, +, .)$ is a ring.]
A subset S of R is called a **subring** of R iff the system $(S, +, .)$ is a ring, i.e. iff S is a ring with respect to the addition and multiplication which are defined on R and therefore on S (see remark (c) in § 27).

Definition 2. Let F be a field. [This means that the system $(F, +, .)$ is a field.]
A subset K of F is called a **subfield** of F iff the system $(K, +, .)$ is a field.

Familiar illustrations: $\mathbb{Z}$ is a subring of $\mathbb{R}$; $\mathbb{R}$ is a subfield of $\mathbb{C}$.

Notice that, clearly:

64.1 In any ring R, R itself and $\{0\}$ are subrings of R; and, in any field F, F is a subfield of F.

Tests for a subring or subfield

64.2 Let R be a ring and let S be a subset of R. Then S is a subring of R iff the following conditions all hold:

(1) $S \neq \emptyset$;

(2) S is closed under subtraction;

(3) S is closed under multiplication.

Proof. (i) If S is a subring, then the conditions (1), (2), (3) all hold, by virtue, respectively, of the ring axioms A, A, M0 as they apply to S.

(ii) Conversely, suppose that the conditions (1), (2), (3) all hold.

By (1), (2) and the additive version of 36.4, it follows that S is a group with respect to addition; and the group $(S, +)$ is abelian since the addition defined on R is commutative. Hence A holds for S. The condition (3) tells us that M0 holds for S. Finally, M1 and D hold for S since they hold for the whole set R. Therefore $(S, +, .)$ is a ring, i.e. S is a subring of R.

Proof of the stated result is now complete.

Formulation of a test for a subfield of a given field is more subtle. We leave as an exercise proof of the following test.

64.3 Let F be a field and K a subset of F. Then K is a subfield of F iff the following conditions all hold: (1) $K - \{0\} \neq \emptyset$;

(2) K is closed under subtraction;

(3) K is closed under multiplication;

(4) $x \in K - \{0\} \Rightarrow x^{-1} \in K$.

Unity elements in relation to subrings/subfields

Consider the set S of matrices of the form $\begin{bmatrix} x & 0 \\ 0 & 0 \end{bmatrix}$, with $x \in \mathbb{R}$. By means of 64.2, it is easily seen that S is a subring of the ring R of all 2×2 real matrices. Clearly the subring S has a unity, viz. $\begin{bmatrix} 1 & 0 \\ 0 & 0 \end{bmatrix}$; and this unity is not the same as the unity of the whole ring R. The moral is that, in a ring with a unity, the unity of a subring (if it possesses one) need not be the same as the unity of the whole ring.

It is easy to prove that, in a field F, the unity of every subfield must coincide with that of F.

65. Ring homomorphisms

Definition. Let R and R_1 be rings. A **(ring) homomorphism** from R to R_1 means a mapping $\theta : R \rightarrow R_1$ such that

$$\forall x, y \in R, (1)\ \theta(x+y) = \theta(x)+\theta(y) \quad \text{and} \quad (2)\ \theta(xy) = \theta(x)\theta(y).$$

Thus, as in group theory, a homomorphism is a mapping which, so to speak, respects the binary operations in the systems concerned.

Let $\theta : R \rightarrow R_1$ be a ring homomorphism. If we temporarily forget the multiplication defined on R and R_1, we see that θ is a group homomorphism from the additive group $(R, +)$ to the additive group $(R_1, +)$. By 47.1 and 47.2, it follows that:

65.1 If $\theta : R \rightarrow R_1$ is a ring homomorphism (R, R_1 being rings), then
  (i) $\theta(0) = 0_1$ (where $0, 0_1$ are the zeros of R, R_1, respectively),
  (ii) $\forall x \in R,\ \theta(-x) = -\{\theta(x)\}$.

In the same vein, we note that the kernel of a ring homomorphism $\theta : R \rightarrow R_1$ already has a meaning, viz.

$$\ker \theta = \{x \in R : \theta(x) = 0_1\} \quad \text{(where } 0_1 \text{ is the zero of } R_1);$$

and 47.8 tells us that:

65.2 If $\theta : R \rightarrow R_1$ is a ring homomorphism, then θ is injective iff $\ker \theta = \{0\}$.

To obtain other facts about ring homomorphisms and their kernels, we must bring back into our thinking the multiplication defined on the rings concerned.

65.3 Let $\theta : R \rightarrow R_1$ be a ring homomorphism (R and R_1 being rings). Then:
  (i) for every subring S of R, $\theta(S)$ is a subring of R_1; and, in particular [the case $S = R$] im θ is a subring of R_1;
  (ii) $\ker \theta$ is a subring of R.

  Proof. In each part we use 64.2.
  (i) Let S be a subring of R.
  (1) Since $S \neq \emptyset$, $\theta(S) \neq \emptyset$.
  (2) and (3) Let x, y be arbitrary elements of $\theta(S)$. Then $x = \theta(s)$ and $y = \theta(t)$

for some $s, t \in S$. Hence

$$x - y = \theta(s) - \theta(t)$$
$$= \theta(s) + \theta(-t) \quad \text{(by 65.1(ii))}$$
$$= \theta(s - t) \quad \text{(since } \theta \text{ is a homomorphism).}$$

Since S, being a subring, is closed under $-$, $s - t \in S$. Hence $x - y \in \theta(S)$.

Similarly, $xy = \theta(s)\theta(t) = \theta(st)$, and so, since S is closed under multiplication (so that $st \in S$), $xy \in \theta(S)$.

This proves that $\theta(S)$ is closed under subtraction and multiplication.

Part (i) now follows by 64.2.

(ii) (1) By 65.1(i), $\theta(0) = 0_1$ ($=$ the zero of R_1), i.e. $0 \in \ker \theta$. Therefore $\ker \theta \neq 0$.

(2) and (3) Let $x, y \in \ker \theta$. Then $\theta(x) = \theta(y) = 0_1$. Hence

$$\theta(x - y) = \theta(x) + \theta(-y) \quad \text{(since } \theta \text{ is a homomorphism)}$$
$$= \theta(x) - \theta(y) \quad \text{(by 65.1(ii))}$$
$$= 0_1 - 0_1$$
$$= 0_1$$

and $\theta(xy) = \theta(x)\theta(y) = 0_1 0_1 = 0_1$. So both $x - y$ and xy belong to $\ker \theta$.

This proves that $\ker \theta$ is closed under subtraction and multiplication.

It follows by 64.2 that $\ker \theta$ is a subring of R.

Remarks. (*a*) Ring homomorphisms of special character are given names like those given correspondingly to special group homomorphisms: i.e.

(1) an injective ring homomorphism is called a **monomorphism**;

(2) a surjective ring homomorphism is called an **epimorphism**;

(3) a bijective ring homomorphism is called an **isomorphism**.

(*b*) On reflection, it can be appreciated that the existence of a ring isomorphism from R to R_1 (where R and R_1 are rings) tells us that R and R_1 are, in a significant sense, the same – that they have the same algebraic structure. When a ring isomorphism from the ring R to the ring R_1 exists, we say that R and R_1 are **isomorphic** to each other; and we indicate this by writing $R \cong R_1$.

66. Ideals

We now define a special kind of subring that we call an *ideal*. As indicated in §61, ideals are to ring theory what normal subgroups are to group theory.

Definition. Let S be a subring of the ring R. Then S is called an **ideal** of R iff the condition

$$s \in S \text{ and } x \in R \Rightarrow \text{ both } xs \text{ and } sx \text{ belong to } S$$

holds.

Thus a subring S of the ring R is an ideal of R iff it has a subtle kind of closure property — namely that, when we take an element of the subring S and multiply it (on left or on right) by any element of R (either in S or in $R - S$), the resulting product is always in S.

Simple illustration. In $\mathbb{Z}$, the subring comprising all the even integers is an ideal: for if we multiply an even integer by any integer whatever, we get for the product an even integer.

It is clear from the definition of an ideal that:

66.1 In every ring R, $\{0\}$ and R are ideals of R.

The next result is a simple but useful observation that follows almost immediately from the definition of an ideal.

66.2 Let R be a ring with a one, 1; and let S be an ideal of R. If S contains a unit of R (in particular if $1 \in S$), then S is the whole of R.

Proof. Suppose that S contains the unit u of R (with inverse u^{-1} in R).

Let x be an arbitrary element of R. Since S is an ideal and since $u \in S$ while $xu^{-1} \in R$, it follows from the definition of an ideal that $(xu^{-1})u \in S$, i.e. $x \in S$.

Since x was arbitrary, this proves that S is the whole of R; and the result follows.

A corollary is that a field has no ideals except the obvious ones that 66.1 points out to us, i.e.

66.3 In a field F, the only ideals are F and $\{0\}$.

Proof. If S is an ideal of the field F and $S \neq \{0\}$, then S contains a nonzero element of F, i.e. S contains a unit of F, and hence, by 66.2, $S = F$.

The following result presents the standard simple test for an ideal not already known to be a subring.

66.4 Let R be a ring and let S be a subset of R. Then S is an ideal of R iff the following conditions all hold:
  (1) $S \neq \emptyset$;
  (2) S is closed under subtraction;
  (3) $s \in S$ and $x \in R \Rightarrow xs$ and sx belong to S.

Proof. (i) If S is an ideal of R, then, because S is a subring, the conditions (1) and (2) hold, and because, further, S is an ideal, condition (3) also holds.

(ii) Conversely, suppose that conditions (1), (2), (3) all hold.

Condition (3) tells us that, in particular, S is closed under multiplication; and, by this and conditions (1) and (2), it follows that S is a subring. By condition (3) it now follows that S is an ideal.

Proof of the stated result is now complete.

Worked example. Let R be the ring of all matrices of the form $\begin{bmatrix} x & y \\ 0 & z \end{bmatrix}$ with x, $y, z \in \mathbb{R}$. (Incidentally, one would prove that R is a ring by using 64.2 to prove that R is a subring of the ring of all 2×2 real matrices.) Let S be the subset of R comprising the matrices of the form $\begin{bmatrix} x & y \\ 0 & 0 \end{bmatrix}$ $(x, y \in \mathbb{R})$. Show that S is an ideal of R.

Solution. (1) $S \neq \emptyset$, clearly.

(2) Let $A = \begin{bmatrix} x_1 & y_1 \\ 0 & 0 \end{bmatrix}$, $B = \begin{bmatrix} x_2 & y_2 \\ 0 & 0 \end{bmatrix}$ be two arbitrary members of S $(x_1, x_2, y_1, y_2 \in \mathbb{R})$. Then

$$A - B = \begin{bmatrix} x_1 - x_2 & y_1 - y_2 \\ 0 & 0 \end{bmatrix}, \text{ which belongs to } S.$$

Hence S is closed under subtraction.

(3) Let $A = \begin{bmatrix} x & y \\ 0 & 0 \end{bmatrix}$ be an arbitrary member of S and $X = \begin{bmatrix} a & b \\ 0 & c \end{bmatrix}$ be an arbitrary member of R $(x, y, a, b, c \in \mathbb{R})$. Then

$$AX = \begin{bmatrix} xa & xb + yc \\ 0 & 0 \end{bmatrix} \text{ and } XA = \begin{bmatrix} xa & ya \\ 0 & 0 \end{bmatrix}$$

and both of these matrices belong to S. Thus $A \in S$ and $X \in R \Rightarrow AX$ and XA belong to S.

It follows (by 66.4) that S is an ideal of R.

The next theorem is the first to give substance to the claim that ideals are the analogues in ring theory of normal subgroups in group theory.

66.5 Let $\theta : R \to R_1$ be a ring homomorphism (R, R_1 being rings). Then $\ker \theta$ is an ideal of R.

Proof. By 65.3(ii), $\ker \theta$ is a subring of R.

Let s be an arbitrary element of $\ker \theta$ (so that $\theta(s) = 0_1$, the zero of R_1), and let x be an arbitrary element of R. Then

$$\theta(xs) = \theta(x)\theta(s) = \theta(x).0_1 = 0_1, \text{ and } \theta(sx) = \theta(s)\theta(x) = 0_1.\theta(x) = 0_1,$$

so that both xs and sx belong to $\ker \theta$.

Thus $s \in \ker \theta$ and $x \in R \Rightarrow$ both xs and sx belong to $\ker \theta$.

It follows that $\ker \theta$ is an ideal of R.

Later (in §68), when we construct the ring-theory analogues of quotient groups, it will be apparent that, conversely, every ideal of a ring R is the kernel of at least one homomorphism with domain R.

67. Principal ideals in a commutative ring with a one

Throughout this section R will denote a *commutative* ring with a one.

For each $a \in R$, we denote by (a) the subset $\{ra : r \in R\}$ of R, i.e. the set of all elements obtainable by multiplying the particular element a by an element of R.

67.1 For each $a \in R$, (a) is an ideal of R: moreover $a \in (a)$, and (a) is the smallest ideal of R to which a belongs (i.e. if J is an ideal of R and $a \in J$, then $(a) \subseteq J$).

Proof. Let a be an arbitrary element of R.

We use 66.4 to prove that (a) is an ideal of R.

(1) $(a) \neq \emptyset$ since, for example, $1a \in (a)$, i.e. $a \in (a)$.

(2) Let x, y be arbitrary elements of (a). Then $x = ra$ and $y = sa$ for some r, $s \in R$. Hence $x - y = ra - sa = (r - s)a$, and so $x - y \in (a)$.

This shows that (a) is closed under subtraction.

(3) If s is an element of (a) and $x \in R$, then $s = ra$ for some $r \in R$ and so $xs = x(ra) = (xr)a$, which belongs to (a).

Thus $s \in (a)$ and $x \in R \Rightarrow xs \, (= sx) \in (a)$.

It follows by 66.4 that (a) is an ideal of R; and it has already been remarked (under (1) above) that $a \in (a)$.

Let now J be any ideal of R such that $a \in J$. By the definition of an ideal, $ra \in J$ for every $r \in R$. So every element of (a) belongs to J, i.e. $(a) \subseteq J$.

This shows that (a) is the smallest ideal of R to which a belongs and completes the proof of the theorem.

Nomenclature. For each $a \in R$, the ideal (a) of R is called the **principal ideal of R generated by** a.

Simple illustration. In $\mathbb{Z}$, for each integer m, (m) is the ideal of $\mathbb{Z}$ comprising all the integral multiples of m. As the next theorem shows, such ideals are the only ideals of $\mathbb{Z}$.

67.2 Every ideal of $\mathbb{Z}$ is a principal ideal, i.e. equals (m) for some integer m.

Foreword to proof. A point of some subtlety arises in the proof of 67.2 given below—namely that, for k, $m \in \mathbb{Z}$, the kth multiple of m in the additive group $(\mathbb{Z}, +)$, in the sense of kth "power" of m under the binary operation $+$, is the same as the product km in the ring $\mathbb{Z}$. The student should reflect on this delicate point before proceeding.

Proof of 67.2. Let J be an arbitrary ideal of $\mathbb{Z}$. Then J is a subring of $\mathbb{Z}$ and, in particular, $(J, +)$ is a subgroup of the additive group $(\mathbb{Z}, +)$. But the additive group $(\mathbb{Z}, +)$ is cyclic (1 being a generator) and hence, by 39.4, $(J, +)$ is a cyclic group—generated, say, by m ($m \in \mathbb{Z}$). Therefore (cf. foreword)

$$J = \{km : k \in \mathbb{Z}\} = (m).$$

This proves the stated result.

Definition. An integral domain in which every ideal is a principal ideal is called a **principal ideal domain**.

Thus, since $\mathbb{Z}$ is an integral domain, 67.2 tells us that $\mathbb{Z}$ is a principal ideal domain.

It turns out that every principal ideal domain has nice factorization properties akin to those of $\mathbb{Z}$.

68. Factor rings

Throughout this section R will denote a ring (not necessarily commutative, not necessarily possessing a unity), and J will denote an ideal of R. We are about to carry out the ring theory analogue of the construction of a quotient group.

To begin with, let us temporarily forget that multiplication is defined on R. Then we are left with an additive group $(R, +)$ in which $(J, +)$ is a subgroup—indeed a normal subgroup since $(R, +)$ is an abelian group. By 50.3 the additive quotient group $(R/J, +)$ exists; and it is an abelian group (since, obviously, every quotient group of an abelian group is also abelian). The results of §50, when translated into additive notation, give us certain information about the group $(R/J, +)$, including:

(*a*) each element is a coset of J in R, i.e. an object of the form

$$x + J \ (= \{x + j : j \in J\}) \text{ for some } x \in R;$$

(*b*) the rule of addition of such cosets is

$$(x + J) + (y + J) = (x + y) + J \qquad (x, y \in R);$$

(*c*) in the group $(R/J, +)$ the zero is J and the negative of $x + J$ is $(-x) + J$ (for each $x \in R$).

Taking R/J to mean the set of elements of the group $(R/J, +)$, let us now remember the multiplication on R and propose the rule

$$(x + J)(y + J) = xy + J \qquad (x, y \in R) \tag{1}$$

as a definition of multiplication on R/J. Since each coset has, in general, several labels of the form $x + J$ ($x \in R$), the question of consistency arises here (cf. §18). The question is favourably settled by the following lemma.

68.1 Suppose that $x + J = x' + J (= \xi)$ and $y + J = y' + J (= \eta)$, where $x, x', y, y' \in R$. Then $xy + J$ and $x'y' + J$ (the two meanings that (1) indicates for $\xi\eta$) are equal.

Proof. By the additive version of 42.6, since $x+J = x'+J$, it follows that $x-x' \in J$. Similarly, $y-y' \in J$. Hence, since J is an ideal, $(x-x')y \in J$ and $x'(y-y') \in J$. So, since, in particular, J is closed under addition,

$$(x-x')y + x'(y-y') \in J, \text{ i.e. } xy - x'y' \in J, \text{ i.e. } xy+J = x'y'+J.$$

We can now proceed, knowing that the equation (1) provides a consistent definition of multiplication on the set R/J, so that we have an algebraic system $(R/J, +, .)$.

68.2 (i) The system $(R/J, +, .)$ is a ring (which we henceforth call "the ring R/J").

(ii) The mapping $v: R \to R/J$ given by $x \mapsto x+J$ $(x \in R)$ is a ring epimorphism with kernel J.

(iii) If R is commutative, so is R/J.

(iv) If R has a unity, 1, then, provided $J \subset R$, R/J has a unity, viz. $1+J$.

Proof. (i) We check that the ring axioms hold for $(R/J, +, .)$.

A: Before introducing multiplication, we saw that R/J is an abelian group with respect to addition.

M0: It is clear from the definition of multiplication (equation (1)) that R/J is closed under multiplication.

M1: For all $x, y, z \in R$, we have, using the definition of multiplication on R/J and the fact that the multiplication on R is associative:

$$(x+J)\{(y+J)(z+J)\} = (x+J)(yz+J) = xyz+J$$

and

$$\{(x+J)(y+J)\}(z+J) = (xy+J)(z+J) = xyz+J$$

Hence the multiplication defined on R/J is associative.

D: The distributive laws may be checked using the same approach as has just been used to check M1.

This done, it follows that $(R/J, +, .)$ is a ring.

(ii) It is a formality to verify that v is a ring homomorphism. It is surjective, since the arbitrary element $x+J$ of R/J $(x \in R)$ is the image under v of x. So v is a ring epimorphism. Further, for $x \in R$,

$$x \in \ker v \Leftrightarrow v(x) = \text{zero of } R/J$$
$$\Leftrightarrow x+J = J$$
$$\Leftrightarrow x \in J \quad \text{(by additive version of 42.6)};$$

and hence $\ker v = J$.

(iii) If R is commutative, then, for all $x, y \in R$,

$$(x+J)(y+J) = xy+J = yx+J = (y+J)(x+J)$$

and this shows that R/J is commutative.

(iv) Suppose that R has a unity, 1, and that $J \subset R$.

By 66.2, $1 \notin J$ and so $1 + J \neq J$, i.e. $1 + J$ is a nonzero element of R/J. Moreover, for all $x \in R$,

$$(1 + J)(x + J) = 1x + J = x + J = x1 + J = (x + J)(1 + J)$$

and hence $1 + J$ is a unity in R/J. The final part of the theorem follows.

Nomenclature. The ring R/J is called the **factor ring** of R modulo J. The epimorphism v detailed in 68.2(ii) is called the **natural epimorphism** from R onto R/J.

Illustration of a factor ring. Consider the factor ring $\mathbb{Z}/(m)$, where m is a positive integer. Notice that the typical element $x + (m)$ of $\mathbb{Z}/(m)$ $(x \in \mathbb{Z})$ is what we have previously called C_x, the congruence class of x modulo m: this is so because, for $t \in \mathbb{Z}$,

$$\begin{aligned}
t \in x + (m) &\Leftrightarrow t - x \in (m) \qquad \text{(by additive version of 42.6)} \\
&\Leftrightarrow m|(t - x) \qquad \text{(by definition of (m))} \\
&\Leftrightarrow t \equiv x \,(\text{mod } m) \\
&\Leftrightarrow t \in C_x.
\end{aligned}$$

Thus $\mathbb{Z}/(m)$ is a ring whose elements are the congruence classes modulo m; and in this ring the rules of addition and multiplication are

$$(x + (m)) + (y + (m)) = (x + y) + (m), \quad (x + (m))(y + (m)) = xy + (m),$$

i.e.

$$C_x + C_y = C_{x+y}, \quad C_x C_y = C_{xy}$$

for all $x, y \in \mathbb{Z}$. Thus the factor ring $\mathbb{Z}/(m)$ is nothing other than our old friend $\mathbb{Z}_m$!

Interestingly, therefore, the construction of a general factor ring R/J is a generalization of the construction of the algebraic system $\mathbb{Z}_m$.

We conclude the section by stating the first isomorphism theorem for rings (cf. 52.1).

68.3 Let $\theta : R \to R_1$ be a ring homomorphism (R, R_1 being rings). Then
$$\text{im } \theta \cong R/\text{ker } \theta.$$

The proof is closely analogous to that of 52.1. It consists of establishing that the rule

$$\phi(x + \text{ker } \theta) = \theta(x) \qquad (x \in R)$$

consistently defines a mapping ϕ from $R/\text{ker } \theta$ to R_1 and proving then that ϕ is a ring monomorphism with image set equal to im θ.

69. Characteristic of an integral domain or field

Throughout this section D will denote an integral domain. By 63.1, all that we are about to say will apply in particular when D is a field.

Notice first that, for $x \in D$ and $n \in \mathbb{Z}$, nx already has a meaning, viz. the nth multiple (additive analogue of nth power) of x in the group $(D, +)$. In particular, if $n \in \mathbb{N}$ and $x \in D$, nx means $x + x + \dots$ to n terms. Observe also that, for each $x \in D$, the period of x in the additive group $(D, +)$ is

(i) infinite iff $nx \neq 0$ for all $n \in \mathbb{N}$,

(ii) the positive integer k iff $kx = 0$ while $nx \neq 0$ for $1 \leqslant n \leqslant k-1$.

We focus attention on the period of the unity, 1, of D.

Definition. (1) We say that the integral domain D has **characteristic zero** iff 1 has infinite period in the group $(D, +)$.

(2) We say that D has **characteristic** p ($\in \mathbb{N} - \{1\}$) iff 1 has finite period p in the group $(D, +)$.

Illustrations. (i) The integral domain $\mathbb{Z}$ has characteristic zero (since, $\forall n \in \mathbb{N}, n1 \neq 0$).

(ii) The field $\mathbb{Z}_5$ has characteristic 5. Its unity is C_1 and, for $n \in \mathbb{N}$,

$$nC_1 = C_1 + C_1 + \dots \text{(to } n \text{ terms)} = C_n$$

so that nC_1 is zero when $n = 5$ and nonzero when $1 \leqslant n \leqslant 4$.

The following two theorems give basic information about the nonzero characteristic p case and the characteristic zero case, respectively.

69.1 Suppose that the integral domain D has nonzero characteristic p. Then (i) p must be prime, and (ii) p is the period of every nonzero element in the group $(D, +)$.

Proof. (i) As a preliminary, note that, for $m, n \in \mathbb{N}$,

$$
\begin{aligned}
(m1)(n1) &= (1+1+1+ \dots \text{to } m \text{ terms}) . (n1) \\
&= n1 + n1 + n1 + \dots \text{to } m \text{ terms} \qquad \text{(by distributive law)} \\
&= m(n1) \\
&= (mn)1 \qquad \text{(by additive version of 35.10(ii))}.
\end{aligned}
$$

Suppose that p is composite, so that $p = kl$, where $1 < k, l < p$. Then, by the meaning of "characteristic", $p1 = 0$, i.e. $(kl)1 = 0$, i.e. (by the above preliminary) $(k1)(l1) = 0$; and so, since D is an integral domain, $k1 = 0$ or $l1 = 0$. In either case, we have a contradiction, since the pth multiple of 1 is the lowest positive multiple of 1 that equals zero.

It follows that p must be prime.

(ii) Let x be an arbitrary element of $D - \{0\}$. Since

$$px = x + x + \dots \text{to } p \text{ terms} = (1 + 1 + \dots \text{to } p \text{ terms})x = (p1)x = 0x = 0,$$

it follows (cf. 38.4(ii)) that the period of x is a factor of p. Hence, since p is prime, the period of x is 1 or p. But it is not 1, since $x \neq 0$. Therefore the period of x is p. This proves the result.

69.2 Suppose that the integral domain D has characteristic zero. Then, in the group $(D, +)$, every nonzero element has infinite period.

Proof. Let x be an arbitrary element of $D - \{0\}$. For every $n \in \mathbb{N}$, $nx = x + x + \ldots$ to n terms $= (1 + 1 + \ldots$ to n terms$)x = (n1)x$. But $x \neq 0$; and, since D has characteristic zero, $n1 \neq 0$ (for every $n \in \mathbb{N}$). So, since D is an integral domain, it follows that, $\forall n \in \mathbb{N}$, $nx \neq 0$: i.e. x has infinite period in the group $(D, +)$. This proves the result.

Prime subfield of a field

Let F be a field. In the light of the above discussion (and 69.1(i) in particular), we recognize that either (1) F has characteristic zero or (2) F has characteristic p for some prime p. We discuss each of these cases in turn, showing, without going into every detail, that in each case F has an easily picked out subfield isomorphic to a well-known field.

Case 1: F has characteristic zero. In this case, we can define (consistently!) a mapping $\theta : \mathbb{Q} \to F$ by

$$m/n \mapsto (m1)(n1)^{-1} \qquad (m, n \in \mathbb{Z}, n \neq 0)$$

and we can prove that this mapping θ is a ring monomorphism. By the ring theory analogue of 47.6, it follows that im θ is a subfield of F isomorphic to $\mathbb{Q}$. Moreover this subfield is contained in every subfield of F: for, if K is any subfield of F, then, because $1 \in K$, K contains all the elements $k1$ ($k \in \mathbb{Z}$) and hence contains all the elements $(m1)(n1)^{-1}$ ($m, n \in \mathbb{Z}, n \neq 0$), i.e. contains im θ.

Case 2: F has characteristic p, where p is prime. In this case, we can define (consistently!) a mapping $\phi : \mathbb{Z}_p \to F$ by

$$C_n \mapsto n1 \qquad (n \in \mathbb{Z})$$

where C_n denotes the congruence class of n modulo p; and we can prove that this mapping ϕ is a ring monomorphism. It follows that im ϕ is a subfield of F isomorphic to $\mathbb{Z}_p$. Note that the subfield im ϕ consists of the p elements $0, 1, 2.1, \ldots, (p-1)1$. Moreover the subfield im ϕ is contained in every subfield of F: for, if K is any subfield of F, then, because $1 \in K$, it is clear that K must contain all the elements of im ϕ.

Summing up what we have found in the two cases, we have the following theorem.

69.3 In every field F, there is a minimal subfield P (i.e. a subfield P contained in every subfield of F), P being isomorphic to $\mathbb{Q}$ if F has characteristic zero and to $\mathbb{Z}_p$ if F has prime characteristic p.

(*Note*: A field F cannot have more than one such minimal subfield: for, if K, L are both minimal subfields of the field F, then $K \subseteq L$ and $L \subseteq K$, so that $K = L$.)

The unique minimal subfield in a field F is called the **prime subfield** of F.

70. Factorization in an integral domain

Throughout this section, D will denote an arbitrary integral domain.

For $x, y \in D$, the statement "x **divides** y (in D)" will mean simply that $y = tx$ for some $t \in D$. We shall write "$x|y$" as an abbreviation for "x divides y".

Certain fundamental facts can be proved immediately.

70.1 (i) For every $x \in D$, $x|0$.

(ii) For every $x \in D$, $1|x$ and, more generally, every unit in D divides x.

(iii) If $x \ (\in D)$ divides a unit in D, then x is a unit.

(iv) For $x, y \in D$, $x|y$ and $y|x$ are both true iff x, y are unit multiples of each other (i.e. iff $y = ux$ for some unit u of D and correspondingly $x = u^{-1}y$).

Proof. (i) follows from the fact that $0 = 0 . x$ (for all $x \in D$).

(ii) Let $x \in D$, and let u be a unit in D. Then the equation $x = u(u^{-1}x)$ shows that $u|x$.

(iii) If $x \ (\in D)$ divides u, u being a unit in D, then $u = tx$ for some $t \in D$, and hence

$$1 = u^{-1}u = u^{-1}tx$$

which shows (D being commutative) that x is a unit (with inverse $u^{-1}t$).

(iv) It is clear that if x, y are unit multiples of each other, then $x|y$ and $y|x$.

Conversely, suppose that $x|y$ and $y|x$, so that $y = tx$ and $x = sy$ for some $s, t \in D$.

If x or y is zero, then clearly both are zero so that we have $x = 1y$ and $y = 1x$, showing that x, y are unit multiples of each other.

Henceforth we deal with the remaining case where x and y are nonzero. In this case, because

$$1x = x = sy = s(tx) = (st)x$$

it follows (by 63.4) that $st = 1$; and this shows that s and t are units (each being the inverse of the other). Hence x and y are unit multiples of each other.

The proof is now complete.

Definition (arising from 70.1(iv)). For $x, y \in D$, we say that x is an **associate** of y iff x, y satisfy the equivalent conditions

(1) $x|y$ and $y|x$

(2) x, y are unit multiples of each other.

It is easy to show that:

70.2 The relation $\sim$ (*associateness*) defined on D by

$$x \sim y \Leftrightarrow x \text{ is an associate of } y \qquad (x, y \in D)$$

is an equivalence relation on D. The $\sim$-class of 0 is $\{0\}$; and the units form a single $\sim$-class.

Remarks. (*a*) Because of 70.1 (i, ii, iii) there is not much point in further discussing factorization of zero or factorization of units in D. That is why henceforth we concentrate on factorization of "nonzero non-units" in D.

(*b*) Because of 70.1(ii), a factorization of x ($\in D$) of the form $x = uy$, where u is a unit (and where therefore y is an associate of x [since $x = uy$ and $y = u^{-1}x$]) is regarded as *trivial*. And a factorization $x = yz$, which expresses nonzero non-unit x ($\in D$) as the product of factors y and z ($y, z \in D$), is called **non-trivial** iff neither y nor z is a unit.

So, for instance, in the integral domain $\mathbb{Z}$, $6 = 1 \times 6$ and $6 = (-1) \times (-6)$ are trivial factorizations of 6; $6 = 2 \times 3$ and $6 = (-2) \times (-3)$ are non-trivial factorizations of 6.

(*c*) Prompted by the importance of prime numbers in factorization of integers, we say that a nonzero non-unit element x of D is **irreducible** iff x does not have any non-trivial factorization in D. In $\mathbb{Z}$, the irreducible elements are the prime numbers and their negatives.

(*d*) Let x be a nonzero non-unit in D. Any factorization of x in D, say $x = z_1 z_2 \ldots z_m$ ($m \geqslant 2$), can always be tidied into the form

$$x = uy_1 y_2 \ldots y_n \tag{*}$$

where $n \geqslant 1$, u is a unit (perhaps 1) and $y_1, y_2, \ldots, y_n$ are non-units. Such a tidying up is achieved by picking out those of $z_1, \ldots, z_m$ that are units, multiplying them together to form the unit u (taking $u = 1$ if none of $z_1, \ldots, z_m$ is a unit), and re-naming the non-units among $z_1, \ldots, z_m$ as $y_1, \ldots, y_n$. (There must be some non-units among $z_1, \ldots, z_m$, since otherwise $x = z_1 z_2 \ldots z_m$ would be a unit.)

We shall call a factorization of x **tidy** when it is in the form (*) with u a unit (perhaps 1) and each y_i a non-unit.

(*e*) Two tidy factorizations of 6 in $\mathbb{Z}$ are

$$6 = 1 . 2 . 3 \quad \text{and} \quad 6 = (-1) . (-3) . 2.$$

It is reasonable to suggest that these factorizations are not really different, since we can get from one to the other just by juggling around with units (± 1) and changing the order of the factors. We say that they are equivalent factorizations of 6.

More generally, if x is a nonzero non-unit in D, we say that two tidy factorizations

$$x = uy_1 y_2 \ldots y_m, \quad x = vz_1 z_2 \ldots z_n \quad (u, v \text{ units}, y_1, \ldots, z_n \text{ non-units})$$

of x in D are **equivalent** iff $m = n$ and it is possible to pair off the y's and z's so that corresponding y_i and z_j are associates of each other.

(f) Let x be a nonzero non-unit element of D. A **complete** factorization of x in D means a tidy factorization $x = uy_1 y_2 \ldots y_m$ in which all of $y_1, y_2, \ldots, y_m$ are irreducible. For example, a factorization of an integer is complete when the integer is expressed as

$$(\pm 1) \times (\text{a product of primes or their negatives})$$

(g) It is not hard to see that the following is an alternative statement of the fundamental theorem of arithmetic (cf. §13): in $\mathbb{Z}$, every nonzero integer (excepting the units $1, -1$) possesses a complete factorization, and any two such factorizations of the same integer are equivalent.

More generally, we say that D is a **unique factorization domain** iff, for every nonzero non-unit $x \in D$, (i) x possesses a complete factorization and (ii) any two complete factorizations of x are equivalent.

Thus, our re-expressed form of the fundamental theorem of arithmetic says that $\mathbb{Z}$ is a unique factorization domain; and the search for systems with nice factorization properties akin to those of $\mathbb{Z}$ can now be more precisely specified as the search for other unique factorization domains.

A major result in this vein is that every principal ideal domain is a unique factorization domain. There is not space in this textbook to include a proof of this fact, though in the remaining results in this section (especially 70.4) we take some significant steps towards it. (A complete proof of the major result can be found in the textbook by Hartley and Hawkes [7].)

The following theorem shows how statements about factors, etc., correspond to statements about principal ideals.

70.3 For $x, y \in D$ (D being an integral domain but not necessarily a principal ideal domain):

(i) $x|y \Leftrightarrow (y) \subseteq (x)$;

(ii) x is a unit $\Leftrightarrow (x)$ is the whole of D;

(iii) x, y are associates of each other $\Leftrightarrow (x) = (y)$.

Proof. (i) For $x, y \in D$,

$x|y \Rightarrow y = tx$ for some $t \in D$

$\quad \Rightarrow y \in (x)$

$\quad \Rightarrow (y) \subseteq (x) \qquad$ (since (see 67.1) (y) is the smallest ideal to which y belongs) and, conversely,

$$(y) \subseteq (x) \Rightarrow y \in (x) \quad \text{(since } y \in (y))$$
$$\Rightarrow y = tx \text{ for some } t \in D$$
$$\Rightarrow x | y$$

(ii) For $x \in D$,

$$x \text{ is a unit} \Rightarrow (x) \text{ contains a unit}$$
$$\Rightarrow (x) = D \quad \text{(by 66.2)}$$

and, conversely,

$$(x) = D \Rightarrow 1 \in (x)$$
$$\Rightarrow 1 = tx \text{ for some } t \in D$$
$$\Rightarrow x \text{ is a unit.}$$

(iii) For $x, y \in D$,

$$x, y \text{ are associates of each other} \Leftrightarrow x | y \text{ and } y | x \quad \text{(by definition)}$$
$$\Leftrightarrow (y) \subseteq (x) \text{ and } (x) \subseteq (y) \quad \text{(by (i))}$$
$$\Leftrightarrow (x) = (y)$$

The final result in this section (70.4) is a generalization of the theorem 12.3 about highest common factors of integers, which played such a key role in our discussion of factorization of integers in chapter 2. For this purpose we make the following definitions.

Definitions. If $a_1, a_2, \ldots, a_n$ are elements of the integral domain D, then (1) an element x of D is called a **common factor** of $a_1, a_2, \ldots, a_n$ iff $x | a_i$ for $i = 1, 2, \ldots, n$, and (2) an element x of D is called a **highest common factor** of $a_1, a_2, \ldots, a_n$ iff x is a common factor of $a_1, a_2, \ldots, a_n$ and every common factor of these elements of D divides x (cf. 12.4 and the remarks following it).

Here now is the generalization of 12.3 to an arbitrary principal ideal domain.

70.4 Let $a_1, a_2, \ldots, a_n$ be nonzero elements of the principal ideal domain D. Then:

(i) $a_1, a_2, \ldots, a_n$ have a highest common factor;
(ii) any two such highest common factors are associates of each other;
(iii) each such highest common factor is expressible in D as a linear combination of $a_1, a_2, \ldots, a_n$, i.e. in the form

$$d_1 a_1 + d_2 a_2 + \ldots + d_n a_n, \quad \text{with} \quad d_1, d_2, \ldots, d_n \in D.$$

Proof. Let J be the set of all linear combinations in D of $a_1, a_2, \ldots, a_n$. By use of 66.4, it is easy to show that J is an ideal of D. Since D is a principal ideal domain, it follows that $J = (x)$ for some $x \in D$.

(i) Since each a_i is expressible as a linear combination of $a_1, a_2, \ldots, a_n$ $(a_i = 0a_1 + \ldots + 1a_i + \ldots + 0a_n)$, it follows that each $a_i \in J$, i.e. that each $a_i = t_i x$ for some $t_i \in D$, i.e. that x divides each a_i. Thus x is a common factor of $a_1, a_2, \ldots, a_n$. Moreover, since $x \in (x) = J$,

$$x = c_1 a_1 + c_2 a_2 + \ldots + c_n a_n \text{ for some } c_1, c_2, \ldots, c_n \in D,$$

and from this last equation it is evident that every common factor of $a_1, a_2, \ldots, a_n$ divides x. Hence x is a highest common factor of $a_1, a_2, \ldots, a_n$. This proves part (i).

(ii) Let y_1, y_2 be any two highest common factors of $a_1, a_2, \ldots, a_n$. Then, by the definition of "highest common factor", $y_1 | y_2$ and $y_2 | y_1$, i.e. y_1, y_2 are associates of each other. This proves part (ii).

(iii) Let y be an arbitrary highest common factor of $a_1, a_2, \ldots, a_n$. Then, by (ii) and the fact proved above that x is a highest common factor, y is an associate of x. Hence, by 70.3(iii), $(y) = (x) = J$. Therefore $y \in J$, i.e. y is expressible as a linear combination in D of $a_1, a_2, \ldots, a_n$. This proves part (iii).

Taking a lead from our work on $\mathbb{Z}$ in sections 12 and 13, the interested student may like to attempt to deduce from 70.4 that, in a principal ideal domain, any two complete factorizations of a given nonzero non-unit are equivalent. There remains, of course, the question of whether every such element in such a ring possesses a complete factorization, to be settled before one can draw the conclusion that every principal ideal domain is a unique factorization domain.

71. Construction of fields as factor rings

In this section R will denote a commutative ring with a one, but not necessarily an integral domain.

Definition. An ideal J of R is called **maximal** iff (1) $J \subset R$ and (2) there is no ideal K of R such that $J \subset K \subset R$.

The title of this section comes from the following theorem.

71.1 In the ring R (commutative with a one), let J be a maximal ideal. Then R/J is a field.

Proof. Since $J \subset R$, it follows from 68.2(iii, iv) that R/J is a commutative ring with a one. It remains to prove that each nonzero element of R/J has a multiplicative inverse in R/J.

Let then $x + J$ $(x \in R)$ be an arbitrary nonzero element of R/J; "$x + J$ nonzero" means that $x + J \neq J$, i.e. $x \notin J$.

Consider the subset

$$K = \{j + rx : j \in J, r \in R\}$$

of R, i.e. the subset of R comprising those elements expressible as the sum of an element of J and a multiple (in R) of x. By using 66.4, it is easy to prove that K is an ideal of R. Since, for each $j \in J$, $j + 0x \in K$, i.e. $j \in K$, it follows that $J \subseteq K$; and since $x = 0 + 1x$, it is also true that $x \in K$. Therefore, since $x \notin J$, K is an ideal such that $J \subset K \subseteq R$. So, since J is a maximal ideal, it follows that $K = R$. Hence $1 \in K$, and thus $1 = j_0 + sx$ for some $j_0 \in J$ and some $s \in R$. From this it follows that $1 - sx \in J$ and hence

$$1 + J = sx + J = (s + J)(x + J).$$

This shows that, in the commutative ring R/J, $x + J$ has $s + J$ as multiplicative inverse.

It now follows that R/J is a field.

Some interesting applications of 71.1 arise through the following lemma.

71.2 Let R be a principal ideal domain, and let p be an irreducible element of R. Then (p) is a maximal ideal of R.

Proof. An irreducible element is, by definition, a non-unit. So, by 70.3(ii), $(p) \subset R$.

Suppose that the ideal (p) of R is not maximal. Then there is an ideal K of R satisfying $(p) \subset K \subset R$. Since R is a principal ideal domain, $K = (x)$ for some $x \in R$; and thus we have $(p) \subset (x) \subset R$. Since $(p) \subset (x)$, $x \mid p$ (by 70.3(i)); i.e. $p = tx$ for some $t \in R$. But p, being irreducible, has no non-trivial factorization in R, and so either x or t is a unit in R, i.e. either x is a unit or x is an associate of p. However, since $(x) \neq R$, x is not a unit (by 70.3(ii)); and, since $(p) \neq (x)$, x is not an associate of p (by 70.3(iii)). Thus we have a contradiction.

It follows that (p) is a maximal ideal of R.

For an illustration of how 71.1 and 71.2 can be used together, consider an arbitrary prime number p. Since p is irreducible in the principal ideal domain $\mathbb{Z}$, (p) is a maximal ideal of $\mathbb{Z}$ (by 71.2); and so, by 71.1, $\mathbb{Z}/(p)$ is a field. Thus we re-discover, by a sophisticated route, that $\mathbb{Z}_p$ is a field (for every prime number p).

To make other applications of 71.2 or of the major theorem (mentioned but not proved in §70) that every principal ideal domain is a unique factorization domain, we must first find other examples of principal ideal domains. This will be done in §73, where we show that polynomial rings of a certain important kind are principal ideal domains.

72. Polynomial rings over an integral domain

Although polynomials can be conceived and discussed over rings which are not integral domains, the points of most basic interest arise in connection with polynomials over integral domains and so, in this book, we restrict attention to these.

Throughout this section D will denote an arbitrary integral domain.

Suppose that R is a commutative ring containing D as a subring and that x is an element of R. Then each element of R expressible in the form

$$a_0 + a_1 x + a_2 x^2 + \ldots + a_n x^n$$

where $n \geqslant 0$ and each $a_i \in D$, is called a **polynomial in** x **over** D. It is a straightforward application of 64.2 to prove that:

72.1 (With D and R as above), for each $x \in R$, the set of polynomials in x over D is a subring of R.

The subring of R consisting of all the polynomials in x over D is denoted by $D[x]$. Notice that $D[x]$ contains D as a subring and x as an element. Moreover, fairly obviously, $D[x]$ is the smallest subring of R with these properties.

Illustration. $\mathbb{Z}[\sqrt{2}]$ is the subring of $\mathbb{R}$ comprising all numbers $a + b\sqrt{2}$ with $a, b \in \mathbb{Z}$: for every expression of the form

$$a_0 + a_1 \sqrt{2} + a_2 (\sqrt{2})^2 + \ldots + a_n (\sqrt{2})^n \qquad (n \geqslant 0, \text{ each } a_i \in \mathbb{Z})$$

can be simplified to $a + b\sqrt{2}$, with $a, b \in \mathbb{Z}$.

As in the previous general discussion, let x be an element of the commutative ring R, where $D \subseteq R$. In the following definitions we make a distinction between two possibilities for the standing of the element x relative to D—a distinction of great significance for the structure of $D[x]$.

Definitions. (1) We say that x is **transcendental** over D iff, for $a_0, a_1, \ldots, a_n \in D$ (n being any nonnegative integer whatever),

$$a_0 + a_1 x + a_2 x^2 + \ldots + a_n x^n = 0 \Rightarrow a_0 = a_1 = a_2 = \ldots = a_n = 0$$

(i.e. x is transcendental over D iff there is no non-trivial way to express 0 as a polynomial in x over D.)

(2) We say that x is **algebraic** over D iff x is not transcendental over D, i.e. iff it is possible in a non-trivial way to express 0 as a polynomial in x over D.

For example, $\sqrt{2}$ is algebraic over $\mathbb{Z}$, because, as is shown by the equation $0 = 2 + 0\alpha - \alpha^2$ (where $\alpha = \sqrt{2}$), it is possible to express 0 as a polynomial in $\sqrt{2}$ over $\mathbb{Z}$ in a non-trivial way. On the other hand, it is known that the important real numbers π and e are transcendental over $\mathbb{Z}$. Proof of this may be found in the famous textbook by Hardy and Wright [8].

In this section we are mostly concerned with rings $D[x]$ where x is transcendental over D, and we now consider some results and definitions relating to such rings.

72.2 Suppose that x is transcendental over D. Then each nonzero member of

$D[x]$ can be expressed in just one way in the form

$$a_0 + a_1 x + \ldots + a_n x^n, \text{ with } n \geqslant 0, \text{ each } a_i \in D, \text{ and } a_n \neq 0.$$

Proof. Let f be an arbitrary element of $D[x] - \{0\}$.

Clearly there is at least one way of expressing f in the form indicated. If there were two different such expressions for f, then, by subtracting one from the other, we would succeed in expressing $0 \ (= f - f)$ as a polynomial in x over D in a non-trivial way—a contradiction of the fact that x is transcendental over D.

The result follows.

In the light of 72.2, consider an arbitrary nonzero element f of $D[x]$, where x is transcendental over D. By 72.2, there is just one sequence of coefficients $a_0, a_1, \ldots, a_n$ (with $n \geqslant 0$, $a_n \neq 0$, and each $a_i \in D$) such that $f = a_0 + a_1 x + \ldots + a_n x^n$. In particular, the integer $n \ (\geqslant 0)$ and the coefficients $a_0, a_1, \ldots, a_n$ are uniquely determined by f. We call n the **degree** of f (as a polynomial in x over D), and we call a_n the **leading coefficient** of f. We say that f is **monic** iff its leading coefficient is 1.

Remarks. (a) A polynomial in x of degree 0 over D is a nonzero element of D. Zero, which belongs to $D[x]$, does not have a degree.

(b) If we multiply together two nonzero elements of $D[x]$—say $f = a_0 + a_1 x + \ldots + a_n x^n$ and $g = b_0 + b_1 x + \ldots + b_m x^m$, where a_n and b_m are nonzero—we obtain a product of the form

$$fg = c_0 + c_1 x + \ldots + c_{m+n} x^{m+n}$$

where $c_{m+n} = a_n b_m$, which is nonzero since D is an integral domain. In particular, it is apparent that the product of two nonzero elements of $D[x]$ is also nonzero. Since $D[x]$ is commutative with a one, it follows that:

72.3 If x is transcendental over D, then $D[x]$ is an integral domain.

The next result says essentially that, up to isomorphism, there can only be one ring $D[x]$ with x transcendental over D.

72.4 Let R, R_1 be commutative rings, each containing D as a subring, and let x, x_1 be elements of R, R_1, respectively, each transcendental over D. Then $D[x] \cong D[x_1]$.

Outline of proof. The mapping $\theta : D[x] \to D[x_1]$ given by

$$a_0 + a_1 x + \ldots + a_n x^n \mapsto a_0 + a_1 x_1 + \ldots + a_n x_1^n \qquad (a_0, a_1, \ldots, a_n \in D)$$

is a bijection. The result follows when one has gone through the detail of proving that θ is a ring homomorphism.

Notation. It is conventional to use X to denote an element transcendental over a given integral domain. Accordingly $D[X]$ means the ring (unique up to isomorphism) of polynomials over D in an element which is transcendental over D. (We indicate at the end of the section how it can be proved that, whatever the integral domain D might be, such a ring always exists.) The ring $D[X]$ is often called the ring of **polynomial forms** over D.

To understand why we have taken the above kind of approach to polynomials, it is important to realize that there is a non-trivial distinction between the ring $D[X]$ and a related ring $P(D)$ called the ring of **polynomial functions** over D. The ring $P(D)$ consists of all mappings $f: D \to D$ specifiable by formulae of the form

$$f(x) = a_0 + a_1 x + a_2 x^2 + \ldots + a_n x^n \qquad (x \in D)$$

where $n \geqslant 0$ and the coefficients $a_0, a_1, \ldots, a_n \in D$. Addition and multiplication are defined on $P(D)$ as follows: for $f, g \in P(D)$, $f+g$ and fg are the mappings from D to D given, respectively, by

$$(f+g)(x) = f(x) + g(x), \qquad (fg)(x) = f(x)g(x) \qquad (x \in D).$$

After verifying that $(P(D), +, .)$ actually is a ring, one can show further that the mapping κ from $D[X]$ to $P(D)$ given by

$$a_0 + a_1 X + \ldots + a_n X^n \mapsto f$$

where f is the polynomial function $x \mapsto a_0 + a_1 x + \ldots + a_n x^n$ $(x \in D)$, is a ring epimorphism. But κ is not always an isomorphism, and accordingly one must not presuppose that $D[X] \cong P(D)$, even though, in fact, this is true for many integral domains.

An example showing that $D[X]$ and $P(D)$ may be very different in nature is found in the case $D = \mathbb{Z}_2$. The ring $\mathbb{Z}_2[X]$ is clearly an infinite ring $(1, X, X^2, X^3, \ldots$ all being different members of it); but the ring $P(\mathbb{Z}_2)$ is finite as it is contained in the set of all mappings from $\mathbb{Z}_2$ to $\mathbb{Z}_2$—a set with only 4 members.

Footnote. We end this section by sketching briefly how to prove the important fact that, for every integral domain D, a ring $D[X]$, with X transcendental over D, does exist. (The full details of such a proof are more than slightly long and tedious.)

The underlying idea is that when $D[X]$ does exist, the typical member $a_0 + a_1 X + \ldots + a_n X^n$ $(a_n \neq 0)$ corresponds to (and is completely described by) the coefficient sequence $(a_0, a_1, \ldots, a_n, 0, 0, 0, \ldots)$.

Contrariwise, to construct $D[X]$, we take the set R of all infinite sequences of elements of D that have the property of being zero from some point onwards; and we define addition and multiplication on R in such a way as to mimic the way the coefficient sequences of polynomials behave in addition and multiplication. We then prove that: (1) the system $(R, +, .)$ is now a ring; (2) the sequences $(a_0, 0, 0, 0, \ldots)$ form a subring $\hat{D}$ of R isomorphic to D; (3) the sequence $X = (0, 1, 0, 0, 0, \ldots)$ is transcendental over $\hat{D}$; and (4) $\hat{D}[X]$ is the whole of R. We conclude that we have constructed a ring $\hat{D}[X]$ with $\hat{D} \cong D$. If we now agree to ignore the difference between the element $a_0 \in D$ and the corresponding element $(a_0, 0, 0, 0, \ldots)$ of $\hat{D}$, so that $\hat{D}$ is identified with D, the ring we have constructed is $D[X]$, as required.

73. Some properties of $F[X]$, where F is a field.

Throughout this section F will denote an arbitrary field.

The most interesting properties of $F[X]$ stem from the following division algorithm result.

73.1 Let d be an element of $F[X]$ of degree $n \geqslant 1$. Then, for every $f \in F[X]$, there exist $q, r \in F[X]$ such that

$$f = qd + r \text{ and either } r = 0 \text{ or } r \text{ is nonzero with degree} \leqslant n - 1.$$

Proof. As the equation $0 = 0d + 0$ shows, the theorem is true for the trivial case $f = 0$. So each counterexample to the theorem (if there is one) will have a degree.

Suppose, with a view to obtaining a contradiction, that there exists at least one counterexample. By the well-ordering principle (10.1), it is legitimate to introduce the lowest degree m possessed by a counterexample. Henceforth in the proof let f denote a counterexample of this lowest possible degree m.

Clearly $m \geqslant n$: for, if $m < n$, the equation $f = 0d + f$ would show that f was not a counterexample. Moreover $d \nmid f$ in $F[X]$: for, if $d | f$ in $F[X]$, we would have $f = qd + 0$ for some $q \in F[X]$, showing that f was not a counterexample.

Suppose that $f = a_0 + \ldots + a_m X^m$ and $d = b_0 + \ldots + b_n X^n$ where a_m, b_n are nonzero (and, by the above, $m \geqslant n$). The polynomial $(a_m b_n^{-1} X^{m-n})d$ has the same degree and leading coefficient as f, and hence the difference

$$f_1 = f - (a_m b_n^{-1} X^{m-n})d$$

is a polynomial of degree less than m (a nonzero polynomial, since $d \nmid f$). Since there is no counterexample to the theorem of degree less than m, it follows that $f_1 = q_1 d + r$, for some $q_1, r \in F[X]$, where either $r = 0$ or r is nonzero with degree $\leqslant n - 1$. Hence we have:

$$f = f_1 + (a_m b_n^{-1} X^{m-n})d$$
$$= (q_1 + a_m b_n^{-1} X^{m-n})d + r$$

—an expression of the form in question for f—showing that f is not after all a counterexample.

This is a contradiction, and from it the result follows.

A highly significant corollary is:

73.2 $F[X]$ is a principal ideal domain.

Proof. Let J be an arbitrary ideal of $F[X]$. Our aim is to prove that J must be a principal ideal.

If $J = \{0\}$, then $J = (0)$, which is a principal ideal. And if $J = F[X]$, then $J = (1)$, which is a principal ideal.

Henceforth we deal with the case where $J \neq \{0\}$ and $J \neq F[X]$. In this case J contains a nonzero member; and, by the well-ordering principle, it is legitimate to introduce the lowest degree (n, say) possessed by a nonzero member of J. Note that $n \geqslant 1$: for if $n = 0$, J contains a polynomial of degree 0, i.e. a nonzero element of F, i.e. a unit of $F[X]$, and so (by 66.2) $J = F[X]$, which is not so in this case.

Let d be a polynomial of degree n in J, and let f be an arbitrary member of J. By 73.1, $f = qd + r$, for some $q, r \in F[X]$, where r is either zero or a polynomial of degree $\leqslant n - 1$. Since J is an ideal to which d belongs, $qd \in J$; and since, moreover, $f \in J$, $f - qd \in J$, i.e. $r \in J$. Therefore, since no nonzero member of J has degree less than n while r, if nonzero, has degree $\leqslant n - 1$, it follows that $r = 0$. Hence $f = qd$, and thus $f \in (d)$.

It follows that $J \subseteq (d)$. But, since J is an ideal to which d belongs, $(d) \subseteq J$. Therefore, $J = (d)$.

It has now been shown that every ideal of $F[X]$ is a principal ideal. Since (by 72.3) $F[X]$ is an integral domain, the stated result follows.

Since, as stated in §70, every principal ideal domain is a unique factorization domain, it is a consequence of 73.2 that $F[X]$ is a unique factorization domain—a very helpful fact so far as the handling of polynomials is concerned. Note that in $F[X]$, the units are precisely the nonzero elements of F.

It also follows from 73.2, by the theorems of §71, that:

73.3 If f is an irreducible element of $F[X]$, then $F[X]/(f)$ is a field.

As an illustration of 73.3, let us consider in detail the case $\mathbb{R}[X]/(p)$, where $p = X^2 + 1$. We shall take it as read that $X^2 + 1$ is an irreducible element in $\mathbb{R}[X]$: so, by 73.3, $\mathbb{R}[X]/(p)$ is a field.

Let v be the natural epimorphism from $\mathbb{R}[X]$ onto $\mathbb{R}[X]/(p)$. Then (see 68.2(ii)) $\ker v = (p)$, and hence

$$\ker(v|\mathbb{R}) = \mathbb{R} \cap (p) = \{0\}$$

It follows that $v|\mathbb{R}$ is a monomorphism from $\mathbb{R}$ to the field $\mathbb{R}[X]/(p)$, and thus $v(\mathbb{R})$ is an isomorphic copy of $\mathbb{R}$ inside $\mathbb{R}[X]/(p)$. From now on, let us identify this copy of $\mathbb{R}$ with $\mathbb{R}$ itself by ignoring the difference between $v(x)$ and x when $x \in \mathbb{R}$. This gives us:

(A) $\mathbb{R} \subseteq \mathbb{R}[X]/(p)$.

Let f be an arbitrary element of $\mathbb{R}[X]$. By 73.1, $f = qp + r$ for some q, $r \in \mathbb{R}[X]$ with $r = a + bX$ for some $a, b \in \mathbb{R}$. Hence

$$\begin{aligned} v(f) &= v(q)v(p) + v(r) \qquad \text{(since } v \text{ is a ring homomorphism)} \\ &= v(r) \qquad \text{(since } p \in \ker v) \\ &= v(a) + v(b)v(X) \\ &= a + bv(X) \qquad \text{(by identification of } v(\mathbb{R}) \text{ with } \mathbb{R}) \\ &= a + bi \end{aligned}$$

where $i = v(X)$. Since v is surjective, we deduce that:

(B) every element of $\mathbb{R}[X]/(p)$ is expressible in the form $a + bi$ for some a, $b \in \mathbb{R}$.

Since $X^2 + 1 \in \ker v$, we have

$$0 = v(X^2 + 1) = (v(X))^2 + v(1) = i^2 + 1$$

so that:

(C) in $\mathbb{R}[X]/(p)$, $i^2 = -1$.

By using (C) or otherwise, one can easily improve (B) to:

(D) every element of $\mathbb{R}[X]/(p)$ is expressible in exactly one way in the form $a + bi$, with $a, b \in \mathbb{R}$.

By (A), (C), (D), we recognize that the field $\mathbb{R}[X]/(p)$ is none other than the field of complex numbers.

Thus, leaving aside the detail of the above analysis, we see that 73.3 enables one to perform one of the most important steps in elementary mathematics, viz. the construction of the complex field from the real field. More pleasing still is the fact that, as detailed investigation proves, 73.3 can be used to do something more general—to construct in an economical way from a given field F, not containing any root of a certain polynomial equation, a new field K that contains F as a subfield and does contain a root of the equation. So 73.3 points forward to further important explorations in the theory of fields and polynomial equations, and at the same time offers a more satisfying understanding than was hitherto possible of the construction of $\mathbb{C}$ from $\mathbb{R}$ by revealing the generality of which that construction is a particular manifestation.

EXERCISES ON CHAPTER TEN

1. Suppose that the ring R has the property that $x^2 = x$ for every $x \in R$. Prove that $\forall x \in R$, $-x = x$; and show further that R is commutative.

2. Let a be a fixed element of a ring R, and let $C(a)$ be the subset $\{x \in R : xa = ax\}$. Prove that $C(a)$ is a subring of R.

3. Let R be a ring. A *left unity* in R means a nonzero element $e \in R$ satisfying $ex = x$ for all $x \in R$; and similarly a *right unity* in R means a nonzero element $f \in R$ satisfying $xf = x$ for all $x \in R$.

(i) Suppose that R contains one and only one left unity e. By considering $x + e - xe$, where x is an arbitrary element of R, prove that e is in fact a unity in R (i.e. both a left unity and a right unity).

(ii) Suppose instead that R satisfies the postulate

$$xy = 0 \Rightarrow x = 0 \text{ or } y = 0 \qquad (x, y \in R).$$

Could R contain a left unity which is not a right unity?

4. A *Gaussian integer* is a complex number of the form $m + in$, with m, n integers. Use 64.2 to prove that the set of all Gaussian integers is a ring. Prove also that in this ring the group of units is $\{1, -1, i, -i\}$. [*Hint*: $\forall z_1, z_2 \in \mathbb{C}, |z_1 z_2|^2 = |z_1|^2 |z_2|^2$; and $|z|^2$ is a non-negative integer for every Gaussian integer z.]

5. Let a be a fixed element in the commutative ring R, and let $A = \{x \in R : xa = 0\}$. Prove that A is an ideal of R.

6. Let R be a ring. The centre of R is the subset $Z(R)$ defined by

$$Z(R) = \{x \in R : xr = rx \text{ for all } r \in R\}.$$

Show that $Z(R)$ is a subring of R, but not an ideal of R if R is a non-commutative ring with a one.

7. Let J and K be ideals of a ring R.

(i) Prove that $J \cap K$ is also an ideal of R.

(ii) Let $J + K = \{j + k : j \in J, k \in K\}$. Prove that $J + K$ is also an ideal of R and is the smallest ideal of R that contains both J and K as subsets.

(iii) Show that if neither J nor K is $\{0\}$ and if R is an integral domain, then $J \cap K \neq \{0\}$.

8. (i) Let R be a commutative ring with a one in which the only ideals are $\{0\}$ and R. Prove that R is a field.

(ii) By considering the ring of all 2×2 real matrices, show that a non-commutative ring R with a one but with no ideals except $\{0\}$ and R need not be a division ring.

9. Let D be an integral domain, and let a be a nonzero element of D which is not a unit. Show that, $\forall n \in \mathbb{N}, (a^{n+1}) \subset (a^n)$.

Deduce that (i) in an integral domain which is not a field there are infinitely many different ideals, (ii) every finite integral domain is a field.

10. Let J be an ideal of the ring R. Show that the ring R/J is commutative iff $xy - yx \in J$ for every $x, y \in R$.

Deduce that if K_1 and K_2 are ideals of R and both R/K_1 and R/K_2 are commutative, then $R/(K_1 \cap K_2)$ is also commutative.

11. Let $\theta : F \to R$ be a ring homomorphism, where R is a ring and F is a field. Prove that either θ is the "zero mapping" (i.e. $\theta(x) = 0$ for all $x \in F$) or else θ is a monomorphism.

12. Let $\theta: R \to R_1$ be a ring homomorphism, where R, R_1 are rings with unities 1, 1_1, respectively, and where R_1 is an integral domain. Suppose that $\ker \theta \neq R$. Show that $\theta(1) = 1_1$.

13. Let R, S be as in the worked example in §66. Give an alternative proof that S is an ideal of R by showing that the mapping from R to $\mathbb{R}$ given by $\begin{bmatrix} x & y \\ 0 & z \end{bmatrix} \mapsto z$ is a ring homomorphism with kernel S. Deduce the additional information that $R/S \cong \mathbb{R}$.

14. Let R be a commutative ring and J an ideal of R.

(i) Let $a, b \in R$, and suppose that a^m and b^n belong to J for certain positive integers m, n. Taking it as read that the binomial theorem (for a positive integral power) is applicable in R, show that $(a - b)^{m+n-1} \in J$.

(ii) Let $K = \{x \in R : x^m \in J \text{ for some } m \in \mathbb{N}\}$. Prove that K is an ideal of R.

15. Let R denote a ring with a one. An element x of R is termed *nilpotent* iff $\exists n \in \mathbb{N}$ s.t. $x^n = 0$.

(i) Give an example of a nonzero nilpotent element in $\mathbb{Z}_{12}$. Show that if p and q are distinct prime numbers, then $\mathbb{Z}_{pq}$ contains no nonzero nilpotent elements.

(ii) Show that if J is an ideal of R and x is a nilpotent element of R, then $x + J$ is a nilpotent element of R/J.

(iii) Show that if K is an ideal of R and if all elements of K are nilpotent and all elements of the ring R/K are nilpotent, then all elements of R are nilpotent.

(iv) Show that if x is a nilpotent element of R, then $1 - x$ is a unit in R.

In the remaining parts of the exercise, *suppose that R is commutative.*

(v) Using part (ii) of exercise 14, show that the set N of all nilpotent elements in R is an ideal of R and that R/N contains no nonzero nilpotent elements.

(vi) Using part (ii), show that N is contained in every maximal ideal of R.

16. Let D be an integral domain. Show that

$$x^2 = 1 \Rightarrow x = 1 \text{ or } x = -1 \qquad (x \in D).$$

Deduce that if D contains only finitely many units, then the product of these units equals -1. Hence prove that, for every prime number p, $(p-1)! \equiv -1 \pmod{p}$. [This is part of the result known as *Wilson's theorem* in Number Theory.]

17. Let R be a commutative ring with a one. An ideal of R is called a *prime* ideal iff the complement in R of the ideal is non-empty and closed under multiplication. Prove, for J an ideal of R, that

$$R/J \text{ is an integral domain} \Leftrightarrow J \text{ is a prime ideal.}$$

Deduce that every maximal ideal of R is prime. Deduce also that in $\mathbb{Z}$ the prime ideals are $\{0\}$ and the ideals (p) with p prime.

18. Let D be an integral domain. A nonzero non-unit element $p \in D$ is called a *prime* element iff it satisfies the condition

$$p|ab \Rightarrow p|a \text{ or } p|b \qquad (a, b \in D).$$

Prove that (i) every prime element of D is irreducible; (ii) for nonzero $q \in D$, the ideal (q) of D is a prime ideal of D iff q is a prime element. (See exercise 17 for the meaning of "prime ideal".)

Deduce that, *if D is a principal ideal domain,* then (1) every nonzero prime ideal of D is a maximal ideal, and (2) every irreducible element of D is prime (so that "prime element" and "irreducible element" are synonymous in a principal ideal domain).

19. Let F be a finite field of characteristic p (p prime), and let σ be the mapping from F to F given by $\sigma(x) = x^p$. Prove that σ is an automorphism of F (i.e. an isomorphism from F to itself).

20. Let F be a field. Prove that the groups $(F, +)$ and $(F - \{0\}, .)$ are not isomorphic.

21. Let R be a ring. Let $J_1, J_2, J_3, \ldots$ be an infinite sequence of ideals of R such that $J_n \subseteq J_{n+1}$ for all $n \in \mathbb{N}$. Let K be the union of $J_1, J_2, J_3, \ldots$. Prove that K is an ideal of R.

Deduce that in a principal ideal domain there cannot exist an infinite sequence of ideals $J_1, J_2, J_3, \ldots$ with $J_n \subset J_{n+1}$ for all $n \in \mathbb{N}$.

22. Prove the converse of 71.1, viz. that if J is an ideal of R, R being a commutative ring with a one, and R/J is a field, then J is a maximal ideal of R.

23. Let $\theta = i\sqrt{5}$, and let $R = \mathbb{Z}[\theta] = \{a + b\theta : a, b \in \mathbb{Z}\}$. For $\alpha = a + b\theta \in R$ ($a, b \in \mathbb{Z}$), we call the integer $|\alpha|^2 (= a^2 + 5b^2)$ the *norm* of α, and we write $N(\alpha)$ for the norm of α. Note that, by elementary properties of complex numbers, $N(\alpha\beta) = N(\alpha)N(\beta)$ for all $\alpha, \beta \in R$.

Prove that: (i) the units in R are 1 and -1, these being the only elements of R with norm 1; (ii) there is no element of R with norm equal to 2 or 3; (iii) 2, 3, $1 + i\sqrt{5}$, $1 - i\sqrt{5}$ are irreducible elements of R; (iv) R is not a unique factorization domain.

Let $J = \{a + b\theta : a, b \in \mathbb{Z}, a \equiv b \pmod{2}\}$. Prove that J is an ideal of R, but not a principal ideal.

If K is the ideal (2) of R, is R/K an integral domain?

24. Let D be an integral domain. Prove that $D[X]/(X) \cong D$. Deduce that if D is not a field, then $D[X]$ is not a principal ideal domain.

25. When an element of $\mathbb{Z}[X]$ is expressed as $a_0 + a_1 X + \ldots + a_n X^n$ ($a_0, a_1, \ldots, a_n \in \mathbb{Z}$), we call a_0 the "constant term". Prove that the set of polynomials in $\mathbb{Z}[X]$ with even constant term is an ideal of $\mathbb{Z}[X]$, and hence show that $\mathbb{Z}[X]$ is not a principal ideal domain. [Nevertheless $\mathbb{Z}[X]$ is a unique factorization domain.]

26. Let J be a nonzero ideal in $F[X]$, F being a field. Prove that there is exactly one monic polynomial $f \in F[X]$ such that $J = (f)$.

27. Prove that the ring of Gaussian integers (see exercise 4) is a principal ideal domain.

28. In elementary algebra one meets a theorem to the effect that "a polynomial has α as a root (or zero) iff $X - \alpha$ is a factor of the polynomial". Give a precise statement and proof of this theorem, distinguishing carefully between a polynomial f and the corresponding polynomial function κf.

Prove that $X^2 + 1$ is an irreducible element of $\mathbb{R}[X]$.

BIBLIOGRAPHY

[1] P. R. Halmos, *Naive Set Theory* (Van Nostrand, 1960, reprinted by Springer-Verlag, 1974).

[2] J. Hunter, D. Monk *et al.*, *Algebra and Number Systems* (Blackie-Chambers, 1971).

[3] B. Baumslag and B. Chandler, *Theory and Problems of Group Theory* (McGraw-Hill, 1968).

[4] I. D. Macdonald, *The Theory of Groups* (Oxford, 1968).

[5] W. R. Scott, *Group Theory* (Prentice-Hall, 1964).

[6] J. S. Rose, *A Course on Group Theory* (Cambridge University Press, 1978).

[7] B. Hartley and T. O. Hawkes, *Rings, Modules and Linear Algebra* (Chapman and Hall, 1970).

[8] G. H. Hardy and E. M. Wright, *An Introduction to the Theory of Numbers*, 4th edition (Oxford, 1960).

APPENDIX TO EXERCISES

Some answers, some hints, some fragments of solutions

Chapter 1 (p. 20)

1. Negations are: (i) $\exists x \in S$ s.t. $x > 3$, (ii) $\forall x \in S$, $x > 3$. (i) is false in case (*a*), true in case (*b*); (ii) is true in both cases. Final answer: (i, *a*), (ii, *a* and *b*).

2. (*a*) (ii); (*b*) (iii); (*c*) (i) and (iii) are true, (ii) and (iv) false.

3. Prove by contradiction. Suppose x/y, y/z, z/x are all equal. Then, since their product is 1, their common value must be 1; and hence $x = y = z$—contradiction.

4. 3, 5, 2, 10, 1.

5. For $x \in S$, $x \in A - B \Leftrightarrow x \in A$ and $x \notin B \Leftrightarrow x \in A \cap \mathscr{C}B$.
$A - (A - B) = A \cap \mathscr{C}(A \cap \mathscr{C}B) = A \cap (\mathscr{C}A \cup B) [7.1] = (A \cap \mathscr{C}A) \cup (A \cap B) [7.4]$
$= \theta \cup (A \cap B) = A \cap B$, etc.

6. Negation is: $\exists x \in A$ s.t. $x \in B$. This is clearly equivalent to $A \cap B \neq \theta$. Hence $A \cap B \neq \theta$ is equivalent to $\neg (A \subseteq \mathscr{C}B)$.

Last part: (i) Suppose $(A \cap X) \cup (B \cap X') = \theta$ for some $X \subseteq S$. Then $A \cap X = \theta$ and $B \cap X' = \theta$. Hence $A \subseteq X'$ and $X' \subseteq B'$. So $A \subseteq B'$, i.e. $A \cap B = \theta$. (ii) Conversely, if $A \cap B = \theta$, then $X = A'$ satisfies $(A \cap X) \cup (B \cap X') = \theta$.

7. (i) Suppose $X \in$ left-side. Then $X \subseteq A$ or $X \subseteq B$. Hence, in any event, $X \subseteq A \cup B$, i.e. $X \in$ right-side. (ii) Prove right-side $\subseteq$ left-side by the approach adopted in part (i); use 7.2(i) to prove left-side $\subseteq$ right-side. Consider $C = [0, 2]$, $D = [2, 4]$; note that $[1, 3] \in \mathscr{P}(C \cup D)$ but ....

8. (*a*) $|A \cup B| = |A| + |B|$. (*b*) $|A \cup B| = |A| + |B| - |A \cap B| < |A| + |B|$.

9. (i) A, $\mathscr{C}_S A$, θ, S. (ii) One method is to show, for $x \in S$, that $x \in$ left-side $\Leftrightarrow x$ belongs to an odd number of A, B, C.

10. Clearly 0 belongs to the intersection. Prove by contradiction that no nonzero number belongs to it. (If $x (\neq 0)$ belongs, then, $\forall n \in \mathbb{N}$, $0 < x \leqslant 1/n$, and so, $\forall n \in \mathbb{N}$, $n \leqslant 1/x$.)

11. $P = X \cup Y$, where $X = A \cap B \cap C$, $Y = A \cap B \cap D$. Clearly $X \subseteq B \cap C \subseteq Q$, and $Y \subseteq A \cap D \subseteq Q$: hence $P \subseteq Q$. If $P = Q$, then $B \cap C \subseteq Q = P \subseteq A$, etc.; and if $B \cap C \subseteq A$ and $A \cap D \subseteq B$, then $X = B \cap C$ and $Y = A \cap D$, etc.

12. Suppose (1) satisfied for some set X. Since $D = (\ldots) \cap B$, $D \subseteq B$; and, since $D = (A \cap B) \cup (X \cap B) [7.4]$, $A \cap B \subseteq D$. Moreover, $A \cup D = A \cup (A \cap B) \cup (X \cap B) = A \cup (X \cap B) = C$. Conversely suppose (2) satisfied, and let $X = D - A$. Then $X \cap B = B \cap D \cap A' = D - A$; and from this it is easy to show (1) satisfied.

13. Since $A \neq \theta$, introduce $a \in A$. Then, if $A \times B = B \times A$, $x \in B \Rightarrow (a, x) \in A \times B \Rightarrow (a, x) \in B \times A \Rightarrow x \in A$, and hence $B \subseteq A$; etc.

14. Prove $(x, y) \in$ left-side $\Leftrightarrow x \in A$ and $y \in B$ and $x \in C$ and $y \in D \Leftrightarrow (x, y) \in$ right-side.

15. Show $(x, y) \in$ left-side $\Rightarrow (x, y) \in$ right-side, and conversely. Using first part, we have: $(A \cup C) \times (B \cup D) = ((A \cup C) \times B) \cup ((A \cup C) \times D) = (A \times B) \cup (C \times B) \cup (A \times D) \cup (C \times D)$, whence second result. In the case $A = B = \{1\}$ and $C = D = \{2\}$, left-side $= \{(1, 1), (2, 2)\}$, while right-side $= \{(1, 1), (2, 2), (1, 2), (2, 1)\}$.

16. (i) 2, (ii) 5, (iii) 15.

Chapter 2 (p. 34)

1. Proof for $x \notin \mathbb{Z}$: Write $x = n + t$, where $n \in \mathbb{Z}$, $t \in (0, 1)$. Then $-x = (-n - 1) + (1 - t)$, and $1 - t \in (0, 1)$. So $[-x] = -n - 1$, while $[x] = n$.

2. $d|c$ and $d|a \Rightarrow d|(a + b)$ and $d|a$ [11.3] $\Rightarrow d|((a + b) - a)$ and $d|a \Rightarrow d|b$ and $d|a \Rightarrow d = 1$; etc.

3. If d is a common factor of $9n + 8$ and $6n + 5$, then $d|(2(9n + 8) - 3(6n + 5))$, i.e. $d|1$.

4. By 12.3, $d = au + bv$ for some $u, v \in \mathbb{Z}$. So if $d|c$, we have $ax + by = c$, where $x = (c/d)u$, $y = (c/d)v$. Converse follows from 11.4.

5. Imitate proof of 12.5.

6. $ud_1 + vd_2 = 1$ for some $u, v \in \mathbb{Z}$. And $a = kd_1 = ld_2$ for some $k, l \in \mathbb{Z}$. Hence $a = (ud_1 + vd_2)a = ud_1 ld_2 + vd_2 kd_1 = (ul + vk)d_1 d_2$.

7. $n \equiv 7 \pmod{12} \Rightarrow n = 12k + 7$ (for some $k \in \mathbb{Z}$) $\Rightarrow n = 4(3k + 1) + 3$, etc.

8. Consider positive integer $a_n a_{n-1} \ldots a_2 a_1 a_0$, meaning $a_n 10^n + a_{n-1} 10^{n-1} + \ldots + a_2 10^2 + a_1 10 + a_0$. Since $10 \equiv 1 \pmod 9$, $10^r \equiv 1 \pmod 9$ for every $r \geq 0$; and hence our integer is congruent (mod 9) to $a_n + \ldots + a_2 + a_1 + a_0$. It is divisible by 9 iff the sum of its digits is divisible by 9.

9. If either $3 \nmid x$ or $3 \nmid y$, then $x^2 + y^2 \equiv 0 + 1$ or $1 + 0$ or $1 + 1 \pmod 3$, and so $3 \nmid (x^2 + y^2)$.

Last part: Show first that if $x = u$, $y = v$, $z = w$ is an integral solution, then all of u, v, w are divisible by 3 and $x = \frac{1}{3}u$, $y = \frac{1}{3}v$, $z = \frac{1}{3}w$ is also an integral solution. If the stated result is false, we can (by 10.1) consider a solution with least possible positive value of $|z|$, and then the observation of the last sentence provides a contradiction.

10. Modulo 11: $3^3 \equiv 5$; hence $3^{3n} \equiv 5^n$; hence $3^{3n+1} (= 3.3^{3n}) \equiv 3.5^n$. Also $2^{4n+3} \equiv 8.5^n$. So $3^{3n+1} + 2^{4n+3} \equiv 11.5^n \equiv 0$.

11. $x = -9$ is the answer one obtains by using Euclid's algorithm.

12. Suppose $a \equiv b \pmod m$. Let $d_1 = \text{hcf}(a, m)$, $d_2 = \text{hcf}(b, m)$. Using "$b = a + km$", one sees that $d_1|b$. Hence d_1 is a common factor of b, m. Hence $d_1 \leq d_2$. Prove similarly that $d_2 \leq d_1$.

13. Consider prime decompositions: integer $m \geq 2$ is a perfect square iff every prime appearing in the decomposition of m does so to an even power.

14. Suppose $\sqrt{n} = u/v$, where $u, v \in \mathbb{N}$. Then $u^2 = nv^2$. Now consider prime decompositions to deduce that n must be a perfect square.

15. Every prime is a factor of $p_1 p_2 \ldots p_n$, so not a factor of $p_1 \ldots p_n + 1$. This contradicts 13.1. *Variant*: suppose that $3, p_1, p_2, \ldots, p_n$ is a complete list of the primes congruent to 3 (mod 4) and obtain a contradiction by considering $4p_1 p_2 \ldots p_n + 3$.

Chapter 3 (p. 46) (Proofs that relations are equivalence relations will not be mentioned.)

1. $x \in C_n \Leftrightarrow [x] = [n] = n \Leftrightarrow x = n + t$ for some $t \in [0, 1)$, etc.

2. (ii) $X \in \alpha$-class of $\emptyset \Leftrightarrow X \cap A = \emptyset \cap A \Leftrightarrow X \cap A = \emptyset \Leftrightarrow X \subseteq T - A \Leftrightarrow X \in \mathcal{P}(T - A)$. (iii) 2^n classes.

3. For example, the relations ρ, σ, τ defined by: $x \rho y \Leftrightarrow |x - y| < 5$; $x \sigma y \Leftrightarrow x \leq y$; $x \tau y \Leftrightarrow x = y = 0$.

4. The ρ-classes are the straight lines of gradient 1.

5. Note that each R-class can be expressed as $\{2^n t : n \geqslant 0\}$, where t is its smallest member. Answer to last part is 9.

7. C_0, C_1, C_3, C_4 all satisfy $x^2 = x$ in $\mathbb{Z}_6$.

8. By Euclid's algorithm, etc., find $1 = 11.97 - 26.41$. So $C_1 = C_{-26} C_{41} = C_{71} C_{41}$: i.e. inverse is C_{71}. Modulo 97: $41x \equiv 2 \Rightarrow C_{41} C_x = C_2 \Rightarrow C_{71} C_{41} C_x = C_{71} C_2 \Rightarrow C_x = C_{142} = C_{45} \Rightarrow x \in C_{45}$; and conversely.

9. 3 β-classes: B_0, B_1 $(= B_4)$, B_2 $(= B_3)$. Definition of multiplication is consistent because $x^2 \equiv (x')^2$ and $y^2 \equiv (y')^2$ (mod 5) $\Rightarrow (xy)^2 \equiv (x'y')^2$ (mod 5). Definition of addition is not consistent: it implies that $B_0 = B_5 = B_1 + B_4 = B_1 + B_1 = B_2$.

Chapter 4 (p. 59)

1. (i) Alleged values of $f(1), \ldots, f(7)$ do not belong to the codomain. (ii) Is $g(0)$ defined? (iii) Contradictory information about $h(0)$. (iv) Definition is inconsistent: $\theta(\frac{1}{2}) = 3$ and $\theta(\frac{2}{4}) = 6$, while $\frac{1}{2} = \frac{2}{4}$.

2. Yes, in both cases.

3. (i) 15. (ii) $[-1, 80]$. (iii) $[-1, \infty)$; no.

4. (i) n^m. (ii) (a) 0, (b) $n(n-1)(n-2) \ldots (n-m+1)$. (c) There are $\binom{n+1}{2}$. n (i.e. $\frac{1}{2} n(n+1) . n$) ways of arranging for two members of S to be mapped to the same element of T, and, for each of these, $(n-1)!$ ways to complete the specification of a surjection. Last answer: $\frac{1}{24} n(3n+1)\{(n+2)!\}$.

5. f: Use 22.1 to prove injective; not surjective, im f being the set of even integers. g: not injective (e.g. $g(1) = g(2)$); surjective since $\forall y \in \mathbb{N}$, $y = g(2y-1)$. Examples: (a) $x \mapsto (x-2)^2 + 1$; (b) $x \mapsto (-1)^x [\frac{1}{2} x]$.

6. For $A \subseteq S$, $f(A) \cup f(S-A) = f(A \cup (S-A)) = f(S) = T$, etc.

7. Let $y \in$ right-side. Then $y = f(a) = f(b)$ for some $a \in A$, $b \in B$. Since f injective, $a = b$; and hence $a \in A \cap B$. So $y \in f(A \cap B)$; etc.

8. "f injective $\Rightarrow \ldots$" follows from question 7. Conversely suppose that $\forall A \subseteq S$, $f(A) \cap f(S-A) = \emptyset$. Let x, y be unequal members of S. Since $f(\{x\}) \cap f(S-\{x\}) = \emptyset$ $f(x) \neq f(y)$, etc.

9. (i) $x \in A \Rightarrow f(x) \in f(A) \Rightarrow x \in B$. (ii) Suppose f injective. Then $x \in B \Rightarrow f(x) = f(a)$ for some $a \in A \Rightarrow x = a$ [since f injective] $\Rightarrow x \in A$. So $B \subseteq A$, and hence, by (i), $A = B$. (iii) $[-3, 3]$.

10. Definition of θ consistent because $C_{(a,b)} = C_{(c,d)} \Rightarrow b - a = d - c$. θ is injective because $b - a = d - c \Rightarrow C_{(a,b)} = C_{(c,d)}$. θ is surjective because $\forall n \in \mathbb{Z}$, $n = \theta(C_{(x,y)})$ where $x = |n| + 1 - n$, $y = |n| + 1$.

11. (i) (a) Number is $|S|$, (b) number is $|T|$. (ii) Take $\beta(x) = C_x$, and define α (consistently!) by $\alpha(C_x) = f(x)$.

12. Let $t = f^{-1}(T)$, so that $f(t) = T$. The definition of T now tells us that if $t \in T$, then $t \notin T$, and conversely—contradiction.

13. (ii) $[1, \infty)$. (iii) $c = 2$. (iii) $f_1^{-1}(y) = 2 - \sqrt{(y-1)}$, $f_2^{-1}(y) = 2 + \sqrt{(y-1)}$.

14. Let y be an arbitrary element of U. Then $y = (g \circ f)(x)$ for some $x \in S$; and so y is the image under g of something, viz. $f(x)$.

15. $\forall x \in S$, $h(f(x)) = (h \circ f)(x) = (h \circ g)(x) = h(g(x))$, and so, since h injective, $f(x) = g(x)$, etc.

16. $h = h \circ i_S = h \circ (f \circ g) = (h \circ f) \circ g = i_S \circ g = g$. Result now follows by 25.1.

17. Let $g_1(x) = 0$ for all $x \in T$; $g_2(x) = 0$ if $x \in$ im f, $g_2(x) = 1$ if $x \in T -$ im f.

18. Let z be an arbitrarily chosen fixed element of S. Define $g: T \to S$ as follows. If $x \in T -$ im f, $g(x) = z$; if $x \in$ im f, $g(x) =$ the unique element $y \in S$ such that $f(y) = x$.

Chapter 5 (p. 68)

1. $\circ$ is associative on $\mathbb{Z}$, since, $\forall x, y, z \in \mathbb{Z}$, $(x \circ y) \circ z = x \circ (y \circ z) = x + y + z - xy - yz - zx + xyz$. $T \neq \emptyset$, and T is closed under $\circ$ since $1 - (x \circ y) = (1 - x)(1 - y)$.

2. Yes: Ω and $\emptyset$, respectively.

3. (i) $\circ$ is associative because, $\forall x, y, z \in S$, $(x \circ y) \circ z = x \circ (y \circ z) = z$. (ii) Suppose that, in semigroup $(S, \ast)$, e and f are two unequal left identities. If there were a right identity g, then $e = e \ast g = g = f \ast g = f$ — contradiction.

4. Positive integers k, l with the stated property exist because, since S finite, the list $a, a^2, a^3, \ldots$ must contain a repetition. From $a^k = a^{k+l}$, by multiplying successively by $a^l, a^{2l}, \ldots$, we deduce $a^k = a^{k+l} = a^{k+2l} = a^{k+3l} = \ldots$. In particular, $a^k = a^{k+kl}$. This gives the result if $l = 1$; otherwise multiply through by $a^{k(l-1)}$.

5. Let $x \in S$, and let $y = xx'$. $y^2 = x(x'x)x' = x(ex') = xx' = y$. Hence $y = ey = (y'y)y = y'y^2 = y'y = e$. Further $xe = x(x'x) = (xx')x = ex = x$.

Chapter 6 (p. 87)

1. Identity is $(1, 1)$. Inverse of (x, y) is (x^{-1}, y^{-1}).

2. $\forall g \in G$, $(\lambda_x \circ \lambda_y)(g) = \lambda_x(\lambda_y(g)) = \lambda_x(yg) = xyg = \lambda_{xy}(g)$, etc.

3. Choose $a \in S$, and define $\lambda_a : S \to S$ by $\lambda_a(x) = ax$ $(x \in S)$. Because of cancellation λ_a is injective and so also surjective (S being finite). Hence $ae = a$ for some $e \in S$. Let x be an arbitrary element of S. Then $aex = ax$ and so (cancelling) $ex = x$. Hence $xex = xx$ and so (cancelling) $xe = x$. Thus e is identity. Existence of inverses now comes quickly from surjectivity of each λ_a. Last part: consider $(\mathbb{N}, .)$.

4. $y^{-2}x^{-2} = (xy)^{-2} = ((xy)^{-1})^2 = (y^{-1}x^{-1})^2$; i.e. $y^{-1}y^{-1}x^{-1}x^{-1} = y^{-1}x^{-1}y^{-1}x^{-1}$. Hence, by cancellation, $y^{-1}x^{-1} = x^{-1}y^{-1}$. Taking inverses now gives $xy = yx$.

5, 6. Straightforward applications of 36.3 or 36.4.

7. (i) $xy = yx \Leftrightarrow (xy)x^{-1} = (yx)x^{-1}$. (ii) Let f be a non-identity element of S_n, where $n \geqslant 3$. Suppose $f(i) = j$ $(\neq i)$. Since $n \geqslant 3$, can introduce element g of S_n interchanging j and k (k different from i, j) and mapping everything else to itself. Find that gfg^{-1} maps i to k so that $gfg^{-1} \neq f$.

8. If $C(x) \subseteq C(y)$, then $x \in C(y)$ (since clearly $x \in C(x)$), and so $xy = yx$. Second part: Suppose $C(x) \subseteq C(y)$. Then $y \in C(x)$ by first part. Further, $\forall t \in C(x)$, $t \in C(y)$ [since $C(x) \subseteq C(y)$] and so $yt = ty$. It follows that $y \in Z(C(x))$. Conversely suppose $y \in Z(C(x))$. Then $yt = ty$ for all $t \in C(x)$; i.e. every element of $C(x)$ commutes with y; i.e. $C(x) \subseteq C(y)$.

9. Can introduce $h \in H - K$ and $k \in K - H$. If $H \cup K$ is a subgroup (so closed under.), then $hk \in H \cup K$, i.e. $hk \in H$ or $hk \in K$, and hence $h^{-1}(hk) \in H$ or $(hk)k^{-1} \in K$ — a contradiction.

10. $H = \mathbb{N}$ in $\mathbb{R} - \{0\}$.

11. Period of A is 6. (Note that $A^3 = -I$.)

12. $x^2 = 1 \Leftrightarrow x^2 x^{-1} = 1x^{-1}$. If all elements of $G - \{1\}$ have period 2, then, $\forall x, y \in G$, $xy = (xy)^{-1} = y^{-1}x^{-1} = yx$.

13. Suppose x has finite period k. Then $(x^{-1})^k = (x^k)^{-1} = 1$, and so period of $x^{-1} \leqslant$ period of x. Reverse inequality is obtained on replacing x by x^{-1}. Deal separately with the case where x has infinite period. *Second part*: Elements of period $\geqslant 3$ can be divided into pairs of the form $\{x, x^{-1}\}$; there remain an even number of elements, one of which is 1.

14. 2, 5, 10.

15. $(xax^{-1})^n = xax^{-1}xax^{-1}xax^{-1} \ldots xax^{-1} = xa1a1a1 \ldots 1ax^{-1} = xa^nx^{-1}$. Hence deduce that, for $n \in \mathbb{N}$, $(xax^{-1})^n = 1 \Leftrightarrow a^n = 1$; etc. (i) Note that $bc = b(cb)b^{-1}$. (ii)

Suppose s, t are distinct elements of period 2 in G. If s, t commute, then st is a third element of period 2; otherwise sts^{-1} is such.

16. 5, 2, 5.   **17.** $-I$ is the only element of period 2.

18. If x has period $2n-1$ ($n \in \mathbb{N}$), then $x = x^{2n} = (x^n)^2$.

19. Prove that C_3 has period 6 by calculating its successive powers.

20. Suppose the contrary. Then there is a generator $x > 1$, and $\ldots, x^{-2}, x^{-1}, 1, x, x^2,$ $x^3, \ldots$ are in ascending order. Hence contradiction on observing that $\frac{1}{2}(1 + x)$ is a member of the group strictly between 1 and x.

21. For $r \in \mathbb{N}$, $(x, y)^r = (1, 1) \Leftrightarrow (x^r, y^r) = (1, 1) \Leftrightarrow x^r = 1$ and $y^r = 1 \Leftrightarrow m|r$ and $n|r$; and the lowest value of r for which this is true is the l.c.m. of m, n. Suppose G_1, G_2 cyclic of relatively prime orders m, n, with generators x, y, respectively. Then (x, y) has period mn, and so $\langle (x, y) \rangle$ is the whole of $G_1 \times G_2$.

22. If $\mathrm{hcf}(r, n) = d > 1$, then $(x^r)^{n/d} = 1$ and so x^r has period less than n. If $\mathrm{hcf}(r, n) = 1$, then, for $s \in \mathbb{N}, (x^r)^s = 1 \Rightarrow n|rs \Rightarrow n|s$, and so x^r has period n. Final questions: 2^{k-1}, 2.

23. If G non-cyclic, then $\langle x \rangle$, where $x \in G - \{1\}$, is a non-trivial proper subgroup of G, etc.

Chapter 7 (p. 102)

1. 4, 5.

2. $HH \subseteq H$ because H is closed under multiplication. Conversely, $h \in H \Rightarrow 1h \in HH \Rightarrow h \in HH$.

3. Let g be an arbitrary element of G. Let $Y = \{x^{-1}g : x \in X\}$. Clearly $|Y| = |X|$; so, because $|X| > \frac{1}{2}|G|$, $X \cap Y \neq \emptyset$. So, for some $x_1, x_2 \in X$, $x_1^{-1}g = x_2$, i.e. $g = x_1 x_2$, etc.

4. (i) $g \in$ left-side $\Rightarrow g = xt$ (where $x \in X$, $t \in Y$ and $t \in Z$) $\Rightarrow g \in XY$ and $g \in XZ$, etc. (ii) If $a^2 \neq 1$, $X(Y \cap Z) = X\emptyset = \emptyset$, while $1 \in XY \cap XZ$. (iii) Suppose $g \in$ right-side. Then $g = xy = xz$ for some $y \in Y$, $z \in Z$. By cancellation $y = z$; and so $g = xy$, where $y \in Y \cap Z$, etc.

5. Use 36.4. After introduction of arbitrary elements $x, y \in gHg^{-1}$, the next step is: $x = gh_1g^{-1}$, $y = gh_2g^{-1}$ for some $h_1, h_2 \in H$. By two applications of 41.2, $|gHg^{-1}| = |Hg^{-1}| = |H|$.

6. (i) Follows from 35.6. (ii) Trivial. (iii) Follows from 36.4. Now suppose H, K subgroups. If HK is a subgroup, then $HK = (HK)^{-1} = K^{-1}H^{-1} = KH$. Conversely, if $HK = KH$, then $HK \neq \emptyset$ and $(HK)(HK)^{-1} = HKK^{-1}H^{-1} = HKKH = HKH = HHK = HK$, etc.

7. (i) $x^{-1}y \in H$. (ii) $ab^{-1} \in H$. (iii) $c \in H$. (iv) $s^{-1}t \in H$.

8. Suppose $xH \cap L \neq \emptyset$. Then there is an element $y \in xH \cap L$. Since $y \in xH$, $xH = yH$; and since $y \in L$ and $H \subseteq L$, $yH \subseteq L$ (as L closed under .), etc.

9. If $c \in aH \cap bK$, then $aH \cap bK = cH \cap cK = c(H \cap K)$.

10. Since $x = x1 \in xH$, we have $x \in yK$. So $yK = xK$, and therefore $xH \subseteq xK$. Hence obviously $x^{-1}(xH) \subseteq x^{-1}(xK)$.

11. $a_i(H \cap K) \neq a_j(H \cap K) \Rightarrow a_i^{-1}a_j \notin H \cap K \Rightarrow a_i^{-1}a_j \notin K$ [all the a's being in the subgroup H] $\Rightarrow a_iK \neq a_jK$. Hence there are at least as many left cosets of K in G as there are left cosets of $H \cap K$ in H. Last part: since $|G:K| = 9$, $|H:H \cap K|$ is finite and $\leqslant 9$. Hence, since $|H \cap K| = 5$, H is finite and $|H| \leqslant 45$. Hence, since $|G:H| = 8$, G is finite and $|G| \leqslant 8 \times 45 = 360$. But, by Lagrange, $|G|$ is divisible by all of $|G:H|$, $|G:K|$, and $|H \cap K|$, so that $|G|$ is at least the l.c.m. of 8, 9, 5, i.e. $|G| \geqslant 360$. It follows that $|G| = 360$.

12. For $h_1, h_2 \in H$, $h_1K = h_2K \Leftrightarrow h_1^{-1}h_2 \in K \Leftrightarrow h_1^{-1}h_2 \in H \cap K$, etc. Note that HK is the union of all the cosets hK with $h \in H$. Unequal such cosets don't overlap; each contains $|K|$ elements; and, by previous part, the number of different such cosets is $|H:H \cap K|$. Hence formula for $|HK|$. *Last part*: note that $|HgK| = |HgKg^{-1}|$, and apply previous part with K replaced by gKg^{-1}.

13. (i) By Lagrange, $|H \cap K|$ divides $|H|$ and $|K|$. (ii) By Lagrange, both $|H|$ and $|K|$ divide $|G|$.

14. $|H \cup K| = |H| + |K| - |H \cap K| = 32 - |H \cap K|$; and, by Lagrange, $|H \cap K|$ is 1 or 2 or 4 or 8. (Can't be 16 because $H \neq K$.).

15. Argue as in worked example 2 of §43 to prove $|G:H| = 1$.

16. $\langle x \rangle \subseteq C(x)$: hence result, by Lagrange.

17. Consider the coset xG_t. Suppose $x(t) = s$. Then $y \in xG_t \Leftrightarrow y^{-1}x \in G_t \Leftrightarrow y^{-1}(x(t)) = t \Leftrightarrow y^{-1}(s) = t \Leftrightarrow y(t) = s$. Hence result.

18. (i) Let x have period k. Then $|G| = qk$ for some $q \in \mathbb{N}$. So $x^{|G|} = (x^k)^q = 1^q = 1$. (ii) By (i), $C_n^{p-1} = C_1$, i.e. $C_{n^{p-1}} = C_1$, i.e. $n^{p-1} \equiv 1 \pmod p$.

19. $|\langle x \rangle \cap \langle y \rangle| = 1$ because $|\langle x \rangle \cap \langle y \rangle|$ divides both $|\langle x \rangle| (= m)$ and $|\langle y \rangle| (= n)$. Suppose $xy = yx$. Clearly $(xy)^{mn} = 1$ and so period of $xy \leqslant mn$. On the other hand suppose $(xy)^r = 1$. Then $x^r = y^{-r} \in \langle x \rangle \cap \langle y \rangle$, i.e. $x^r = y^r = 1$, and so r is divisible by both m and n, etc.

20. Since group non-cyclic, every non-identity element has period 2 or 4. Hence result by 44.4.

21. Each of the subgroups contains 1 and $p-1$ elements of period p (see 44.2); and the m sets of $p-1$ elements so arising are disjoint (see worked example 1 in §43). Hence $m(p-1)$ elements of period p in the subgroups; and hence result (since every element of period p lies in such a subgroup). (i) If a non-cyclic group of order 55 has m subgroups of order 5 and n of order 11, then (from counting the non-identity elements) $4m + 10n = 54$, and hence $m, n \neq 0$. (ii) Let there be m subgroups of order p. Then $m(p-1) = $ number of elements of period $p = p^2 - 1$, i.e. $m = p + 1$. There are 2 more subgroups, viz. $\{1\}$ and the whole group.

22. There are elements x, y of period 3 such that x, x^2, y all different. By 36.5 the subset $\{x^i y^j : 0 \leqslant i,j \leqslant 2\}$ is a subgroup of G. Its order is 9: so $9 \| |G|$. *Generalization*: if G is an abelian group containing more than $p-1$ elements of period p, then $p^2 \| |G|$. By this result, the abelian G of order pq contains $\leqslant p-1$ elements of period p and $\leqslant q-1$ elements of period q. Hence the number of elements of period pq in G is at least $(pq-1) - (p-1) - (q-1) = (p-1)(q-1)$, which is positive.

Chapter 8 (p. 104)

1. θ injective because $\theta(f) = \theta(g) \Rightarrow \alpha \circ f \circ \alpha^{-1} = \alpha \circ g \circ \alpha^{-1} \Rightarrow \alpha^{-1} \circ (\alpha \circ f \circ \alpha^{-1}) \circ \alpha = \alpha^{-1} \circ (\alpha \circ g \circ \alpha^{-1}) \circ \alpha$, etc.; θ surjective because the arbitrary element g of S_Y equals $\theta(\alpha^{-1} \circ g \circ \alpha)$. It is easy to prove that θ is a homomorphism.

2. The mapping $\begin{bmatrix} 1 & t \\ 0 & 1 \end{bmatrix} \mapsto t$ is an isomorphism.

3. (i) $\forall x, y \in G_1$, $(\phi \circ \theta)(xy) = \phi(\theta(xy)) = \phi(\theta(x)\theta(y)) = \phi(\theta(x))\phi(\theta(y))$, etc. (ii) Take arbitrary elements x, $y \in G_1 : x = \theta(s)$, $y = \theta(t)$, where $s = \theta^{-1}(x)$, $t = \theta^{-1}(y)$. Hence $xy = \theta(st)$, and $\theta^{-1}(x)\theta^{-1}(y) = st = \theta^{-1}(xy)$; etc.

4. (i) $y \in \theta(\langle x \rangle) \Leftrightarrow y = \theta(x^n)$ for some $n \in \mathbb{Z} \Leftrightarrow y = \{\theta(x)\}^n$ for some $n \in \mathbb{Z} \Leftrightarrow y \in \langle \theta(x) \rangle$. (ii) $\{\theta(x)\}^k = \theta(x^k) = \theta(1) = 1_1$. Hence period of $\theta(x)$ divides k. If θ is a monomorphism, then, for $r \in \mathbb{N}$, $\{\theta(x)\}^r = 1_1 \Rightarrow x^r \in \ker \theta \Rightarrow x^r = 1 \Rightarrow k|r$, etc.

5. $\ker \theta = \{(1, y) : y \in G_2\}$. $y \mapsto (1, y)$ is an isomorphism from G_2 to $\ker \theta$.

6. Let y be an arbitrary element of $\theta(Z)$ and t an arbitrary element of G_1. $y = \theta(z)$

for some $z \in Z$, and, since θ surjective, $t = \theta(s)$ for some $s \in G$. Hence $yt = \theta(z)\theta(s)$ $= \theta(zs) = \theta(sz) = \theta(s)\theta(z) = ty$; etc.

7. Use 49.3 (along with 36.6).

8. (i) Note that (clearly) $xHx^{-1} = H$ is equivalent to $xH = Hx$ $(x \in G)$. (ii) Only one subgroup of order 2: so it is normal, by (i); subgroups of order 4 are normal by 49.2.

9. (i) Use 36.4, noting that $(k_1 h_1)^{-1}(k_2 h_2) = (h_1^{-1} k_1^{-1} k_2 h_1)(h_1^{-1} h_2)$. (ii) Note that $x(kh)x^{-1} = (xkx^{-1})(xhx^{-1})$.

10. $H \lhd G$ and $K \lhd H$ by 49.2. But $CAC^{-1} = -A \notin K$, and so $K \not\lhd G$.

11. Let $H = \langle x \rangle$. $H \lhd G$, by 49.2. Let $y = x^{2^{n-2}}$, which has period 2. $\forall g \in G$, gyg^{-1} must be an element of H (as $H \lhd G$) with same period as y; etc.

12. (i) Straightforward by 36.3 and definition of normality. (ii) As L, M abelian, $L \cap M \lhd L$ and $L \cap M \lhd M$. Hence L, M are subsets of $N(L \cap M)$, and so $L \cup M \subseteq N(L \cap M)$. Since $N(L \cap M)$ is a subgroup, it follows that $H \subseteq N(L \cap M)$. Hence result, by (i).

13. $\theta(a^m) = \{\theta(a)\}^m = 1_1$. So $a^m \in \ker \theta$. Hence $a^{m|\ker \theta|} = 1$ (cf. exercise 18(i) on chapter 7). Hence result, by 38.4(ii).

14. Clearly always $(xh)^2 h^{-1}(x^{-1})^2 \in H$. For every $x \in G$, $(xH)^2 = x^2 H = H$ (by given property); and so 44.4 applies to G/H.

15. *Last part*: Suppose xH has finite period m in G/H $(x \in G)$. Then $(xH)^m = H$, i.e. $x^m H = H$, i.e. $x^m \in H$. Hence $(x^m)^n = 1$ for some $n \in \mathbb{N}$. So $x \in H$, i.e. $xH = H$.

16. Note that $xy = yx[x, y]$. $[h, k] = h^{-1}(k^{-1}hk) = (h^{-1}k^{-1}h)k$. First version shows $[h, k] \in H$ if $H \lhd G$. Second version shows $[h, k] \in K$ if $K \lhd G$, etc.

17. $[xN, yN] = (xN)^{-1}(yN)^{-1}(xN)(yN) = (x^{-1}y^{-1}xy)N = [x, y]N$. G/N abelian $\Leftrightarrow \forall x, y \in G$, $[xN, yN] = $ identity of G/N, i.e. $[x, y]N = N$, i.e. $[x, y] \in N$. G/H and G/K abelian $\Rightarrow \forall x, y \in G$, $[x, y] \in H$ and $[x, y] \in K \Rightarrow \forall x, y \in G$, $[x, y] \in H \cap K \Rightarrow G/(H \cap K)$ abelian. *Last part*: No (a counterexample is easily obtained by taking G to be a four-group).

18. $yZ = (xZ)^{-1} \Rightarrow (xZ)(yZ) = Z \Rightarrow xyZ = Z \Rightarrow xy \in Z \Rightarrow (xy)x = x(xy) \Rightarrow yx = xy$ (by cancellation of x on the left).

19. Take G_1 to be a four-group and G_2 to be a cyclic group of order 4, with K_1 and K_2 subgroups of order 2.

20. Consider a counterexample G of least order. G is abelian by 44.4. Obtain contradiction by considering order of $G/\langle x \rangle$, where x is any element of $G - \{1\}$.

21. (i) G/H exists by 49.2. And, $\forall x \in G$, $x^2 H = (xH)^2 = H$, since G/H has order 2. (ii) No: consider a subgroup of order 2 in the group of symmetries of the equilateral triangle.

22. (i) $C(x)$ contains $Z(G)$, clearly; and the containment is strict because x is an element of $C(x)$ that does not belong to $Z(G)$. (ii) Let G be a group of order p^2, and let Z be its centre. By Lagrange and 48.4, $|Z| = p$ or p^2. But the former leads to a contradiction, as follows. $|Z|$ being p, we can introduce an element $x \in G - Z$. By part (i) $|C(x)| > p$, and so, since $|C(x)| \mid |G|$, $|C(x)| = p^2$; i.e. $C(x) = G$; i.e. $x \in Z$ —: contradiction.

23. A fairly easy consequence of 49.3, which one could re-phrase by saying that a normal subgroup of G is a subgroup closed under the taking of conjugates in G.

24. For the first part, adapt the proof of 48.4. If $|N| = p$, then, as $|N \cap Z(G)|$ is > 1 and divides $|N|$, $N \cap Z(G)$ must be the whole of N, etc.

25. Let $Z = Z(G)$. (i) By Lagrange, $|Z| = 1$ or p or p^2 or p^3. But not 1 by 48.4; not p^2 by 51.1; and not p^3 since G non-abelian. (ii) Exercise 22(i) enables one to show that, $\forall x \in G - Z$, $|C(x)| = p^2$ and therefore $|\text{cl}(x)| = p$. So the $p^3 - p$ elements of $G - Z$ must partition into $(p^3 - p)/p = p^2 - 1$ conjugacy classes. And there are in addition p

singleton conjugacy classes inside Z. (iii) By exercise 24, every normal subgroup of order p is contained in Z and so (by part (i)) coincides with Z. (iv) By 51.1, G/Z can contain no element of period p^2, and hence all its non-identity elements have period p. So, $\forall x \in G, (xZ)^p = Z$, i.e. $x^p \in Z$.

26. Follow closely the proof of 51.1. Introduce a generator tN of H/N; every element of H then is expressible in the form $t^i n$, with $i \in \mathbb{Z}$ and $n \in N$; etc.

27. (i) By Lagrange, $|Z(G)| = 1$ or p or q or pq. pq is out since G non-abelian; p, q are ruled out by 51.1. (ii) Let x be an element of period p in G. Since $x \notin Z(G)$, $C(x) \subset G$. Hence, by Lagrange and the fact that $p \| C(x)|$, $|C(x)| = p$; hence $|\mathrm{cl}(x)| = q$ (by 48.2). Hence the elements of period p come in batches of q.

28. By exercise 27, a non-abelian group of order 15 would contain $10a$ elements of period 3 and $12b$ elements of period 5, for some $a, b \in \mathbb{N} \cup \{0\}$; hence (from counting non-identity elements) $10a + 12b = 14$—an impossibility. So every group of order 15 is abelian, etc.

29. Consider the mapping from C to G/H given by $c \mapsto cH$. *Second part*: exploit fact that there is only one element of period 2.

30. (i) Apply 52.1 with $\theta|_H$ as the homomorphism. This gives $\theta(H) \cong H/K$ where $K = \ker(\theta|_H)$. The result follows on considering orders. (ii) Let H be an arbitrary subgroup of order m. By part (i), $|v(H)| \mid m$. But also $|v(H)| \mid |G:K|$, by Lagrange, since $v(H)$ is a subgroup of G/K. Hence, as m and $|G:K|$ are relatively prime, $|v(H)| = 1$; i.e. $H \subseteq \ker v = K$. So, since $|H| = |K|$, $H = K$.

31. First part is generalization of worked example 1 of §44. $K \lhd G$ follows from 49.2. Let b be an arbitrary element of $G - K$. By exercise 21, $b^2 \in K$. If b had period p, then (cf. worked example 1 in §43) $K \cap \langle b \rangle$ would be $\{1\}$, in contradiction to $b^2 \in K$. Hence b must have period 2. Since also $ba \in G - K$. $(ba)^2 = 1$; so $bab = (ba)^2 a^{-1} = a^{-1}$.

32. Use 36.3 to prove set is subgroup of group of all permutations of G.

33. θ is a homomorphism because $(\tau_g \circ \tau_h)(x) = \tau_{gh}(x)$ for all g, h, $x \in G$. $g \in \ker \theta \Leftrightarrow \theta(g) = i_G \Leftrightarrow \tau_g(x) = x$ for all $x \in G$, etc. $I = \mathrm{im}\,\theta$; so, by 52.1, $I \cong G/Z(G)$. $I \lhd \mathrm{Aut}\,G$ because, if $\alpha \in \mathrm{Aut}\,G$, $g \in G$, then $\alpha \tau_g \alpha^{-1} = \tau_{\alpha(g)} \in I$.

34. $x^{-1}\alpha(x) = y^{-1}\alpha(y) \Rightarrow \alpha(xy^{-1}) = xy^{-1} \Rightarrow xy^{-1} = 1 \Rightarrow x = y$. Hence the mapping $\beta: G \to G$ given by $\beta(x) = x^{-1}\alpha(x)$ is injective and so surjective (G being finite). Now suppose $\alpha^2 = i_G$. Then, for $g \in G$, $\alpha(g) = \alpha(x^{-1}\alpha(x))$ for some $x \in G = \{\alpha(x)\}^{-1}x = g^{-1}$. G is abelian since, $\forall s, t \in G$, $s^{-1}t^{-1} = \alpha(s)\alpha(t) = \alpha(st) = (st)^{-1} = t^{-1}s^{-1}$, etc. G has odd order, since otherwise (cf. exercise 13 on chapter 6) G would contain an element of period 2 and it would be mapped to itself by α.

35. Let $G = \langle x \rangle$, and let ψ_m denote the mapping from G to G given by $x^i \mapsto x^{mi}$. Key points include: (a) every homomorphism from G to G is equal to ψ_m for some $m \in \mathbb{Z}$; (b) $\psi_m \circ \psi_n = \psi_{mn}$; (c) $\psi_m \in \mathrm{Aut}\,G \Rightarrow x^m$ generates G.

Chapter 9 (p. 000)

1. Second rows are respectively $2\,6\,4\,3\,1\,5$, $3\,4\,6\,2\,1\,5$, $3\,6\,4\,2\,1\,5$, $6\,3\,5\,4\,2\,1$, $1\,3\,5\,4\,2\,6$.

2. $(1\,5\,6\,2\,4\,3)$, $(1\,6)(2\,5\,3)(4)$, respectively.

3. $(1\,2\,5\,4\,3)$, $(1\,4\,2\,3\,5)$, respectively.

4. (a) 2.4^2, $1.2^3.3$. (b) $(1\,3\,8\,5\,4)(6\,9\,10)(2)(7)$; $(1\,4\,7\,9\,10)(2\,3\,8)(5)(6)$; $(1\,4)(2\,6\,8\,9)$ $(3\,7\,5\,10)$; $(1\,5\,7)(2\,6)(3\,9)(8\,10)(4)$; $(2\,8)(3\,5)(6\,9)(7\,10)(1)(4)$; $(1\,4)(2\,6\,8\,9)(3\,7\,5\,10)$; $(1\,5\,3\,4\,8)$ $(2)(6)(7)(9)(10)$. (c) $4, 6, 15$.

5. (a) $(1\,3\,4\,6)(2\,8\,7)(5)$; $(1\,7\,4)(2\,5\,8)(3)(6)$; $(1\,2\,5\,7\,6)(3\,4)(8)$; $(1\,6\,7\,3)(2\,4\,5)(8)$. (b) $12, 3$. (c) $1\,120$. (d) $(1\,4\,7)(2\,8\,5)(3\,6)$.

6. $\begin{pmatrix} 1 & 2 & 3 & 4 & 5 & 6 & 7 \\ 2 & 3 & 5 & 4 & 7 & 1 & 6 \end{pmatrix}$.

7. $2^2.3, 3.4, 2.5$, and 7. Total number is 1 854.

8. (a) 350; (b) 10 640.

9. The elements of period 3 belong to the conjugacy classes consisting of elements of cycle patterns $1^3.3$ and 3^2; both these classes have order 40. The centralizers of $(a\,b\,c)(d)(e)(f)$ and $(a\,b\,c)(d\,e\,f)$ contain respectively $(a\,b\,c)(d\,e)(f)$ and $(a\,d\,b\,e\,c\,f)$.

10. Consider the fact that α^r moves *all* of the objects moved by α r places round a certain circle: if one object gets back to where it started, so do they all.

11. $2\,(1^2.3.4$ and $2.3.4)$.

12. $(1\,2\,4)(3\,9\,7)(5\,8\,6)$.

13. By 57.3, $|{\rm cls}_n(\alpha)| = (n-1)!$ Hence by 48.2 $|C_{S_n}(\alpha)| = n$. Hence result since $\langle \alpha \rangle \subseteq C_{S_n}(\alpha)$ and $\langle \alpha \rangle$ also has order n.

14. $+1$ (6 inverted pairs).

15. $(1\,5\,10\,9\,7)(2\,6)(3\,4)(8), (1\,3\,10\,5\,7\,6)(2\,4)(8\,9)$, respectively; even, odd, respectively.

16. $\sigma^2, \tau^2, \tau^{-1}$.

17. (a) 11; (b) 13.

18. (a) $(1\,4)(2\,5)(3\,6)$; (b) $(1\,2\,3)$.

19. There are 7 conjugacy classes: $\{i\}$, the 45 elements of cycle pattern $1^2.2^2$, the 40 elements of cycle pattern $1^3.3$, the 40 elements of cycle pattern 3^2, the 90 elements of cycle pattern 2.4, and two classes each consisting of 72 elements of cycle pattern 1.5.

20. (a) follows at once from 54.2. (b) By (a) and 59.4, an arbitrary non-identity element of A_n can be expressed as a product of the form $(1\,c_1)(1\,c_2)(1\,c_3)(1\,c_4)\ldots(1\,c_{2r-1})$ $(1\,c_{2r})$, where one may stipulate that adjacent c_js are different since $(1\,c)^2$ simplifies to i; and in this product $(1\,c_{2j-1})(1\,c_{2j})$ can be re-expressed as $(1\,c_{2j}c_{2j-1})$. The case of i is trivial since $i = (1\,2\,3)^3$.

21. $\lambda_c(x) = x \Rightarrow cx = x \Rightarrow c = 1$ — contradiction. So λ_c moves all the objects; hence must have cycle pattern 2^n. Hence, n being odd, it is an odd permutation. So, by 59.6, $\theta(G)$ contains a subgroup of index 2.

22. Otherwise G has order $2n$ with n odd and >1, and by exercise 21 G has a subgroup of index 2, a normal subgroup by 49.2; and this contradicts the simplicity of G.

23. (a) If a, b are arbitrary objects, then (i) i is an element of A_n that maps a to b if $a = b$, and (ii) $(a\,b\,c)$, where c is a third object, is an element of A_n mapping a to b in the case $a \neq b$. (b) Suppose $\sigma \in G$ fixes object a (G being as described in question). Let t be an arbitrary object. Since G transitive, there is an element $\tau \in G$ such that $\tau(a) = t$; and we have: $\sigma(t) = \tau\sigma\tau^{-1}(t)$ [G being abelian] $= \tau\sigma(a) = \tau(a) = t$. Since t arbitrary, it follows that $\sigma = i$. Hence result.

24. (a) is straightforward. (b) Show that $\tau(\in S_n)$ fixes $\sigma(a)$ iff $\sigma^{-1}\tau\sigma \in G_a$. (c) Totalling group element by group element gives that the number of pairs as described is $\sum_{\gamma \in G} F(\gamma)$. Totalling object by object gives that the same total is $\sum_{a=1}^{n} |G_a|$. In this latter sum the terms corresponding to the objects in any particular orbit of size m are m terms each equal to $|G|/m$ (by the result quoted in part (a) of the question); and hence the sub-total corresponding to each orbit is $m \times |G|/m$, i.e. $|G|$. Hence result.

25. (i) A subgroup of order 6 would be normal, by 49.2, and so would be the union of some of the conjugacy classes of A_4 (cf. proof of 60.5). However, this is impossible since the conjugacy classes of A_4 have orders 1, 3, 4, and 4. A_4 is not simple because the four-group formed by i along with the 3 permutations of cycle pattern 2^2 is a non-trivial normal subgroup. (ii) $15 \mid \mid S_5 \mid$; and 15 and $\mid S_5 \mid /15 \,(= 8)$ are relatively prime. However, S_5 contains no subgroup of order 15, since such a subgroup would be cyclic and S_5 contains no element of period 15.

Chapter 10 (p. 152)

1. $\forall x \in R;\ x + x = (x + x)^2 = x^2 + x^2 + x^2 + x^2 = x + x + x + x$, and hence $x + x = 0$. $\forall x,\ y \in R,\ x + y = (x + y)^2 = x^2 + xy + yx + y^2 = x + y + xy + yx$, and so $xy = -yx = yx$.

2. Straightforward by 64.2; note $C(a) \neq \emptyset$ as $0 \in C(a)$.

3. (i) Verify that, for all $y \in R$, $(x + e - xe)y = y$. Hence (e being the unique left unity) $x + e - xe = e$; i.e. $xe = x$ for the arbitrary x. (ii) No, by part (i): for the given postulate is easily shown to preclude the existence of two different left unities (cf. proof of 63.4).

4. If u is a unit (with inverse v), then $1 = |uv|^2 = |u|^2 |v|^2$, and so (since $|u|^2, |v|^2$ integers) $|u|^2 = 1$, etc.

5. Straightforward by 66.4.

6. First part is an easy application of 64.2. If $Z(R)$ were an ideal, then, since $1 \in Z(R)$, $Z(R)$ would be the whole of R—contradiction when R non-commutative.

7. (i, ii) Proofs that the subsets are ideals are straightforward by 66.4. It is clear (since an ideal is closed under $+$) that any ideal containing both J and K must contain every element of $J + K$. (iii) Let a, b be nonzero elements of J, K, respectively. Then $ab \neq 0$ since R is an integral domain; and $ab \in J$ and $ab \in K$ since J and K are ideals.

8. (i) $\forall a \in R - \{0\}$, (a) must be whole of R and so $1 \in (a)$, etc. (ii) Let R be the ring of all 2×2 real matrices (not a division ring), and let J be an ideal of R containing a nonzero matrix A with, say, nonzero (r, s)th entry λ. Let E_{ij} be the 2×2 matrix with (i, j)th entry 1 and all other entries zero. Then $(1/\lambda)E_{ir}AE_{sj} = E_{ij}$, and $(\forall i, j)$ this belongs to the ideal J. Now it is easy to prove that J is whole of R.

9. $(a^{n+1}) \subseteq (a^n)$ is trivial. $(a^{n+1}) = (a^n) \Rightarrow a^n \in (a^{n+1}) \Rightarrow a^n = a^{n+1}x$ for some $x \in D \Rightarrow 1 = ax$ [63.4]—contradiction, a being a non-unit. (i) There is a non-unit a, and $(a), (a^2), (a^3), \ldots$ are different ideals. (ii) Follows immediately from (i).

10. R/J commutative $\Leftrightarrow \forall x,\ y \in R,\ (x + J)(y + J) = (y + J)(x + J) \Leftrightarrow \forall x,\ y \in R,\ xy + J = yx + J$, etc. Second part. R/K_1 and R/K_2 both commutative $\Rightarrow \forall x,\ y \in R,\ xy - yx \in K_1$ and K_2, etc.

11. $\ker \theta$ is an ideal of F: hence result by 66.3.

12. If $\theta(1)$ were 0, then, $\forall x \in R,\ \theta(x) = \theta(x1) = \theta(x)\theta(1) = 0$—contradiction. So $\theta(1) \neq 0$. Result follows now on noting that $1_1\theta(1) = \theta(1) = \theta(1^2) = \theta(1)\theta(1)$.

13. $R/S \cong \mathbb{R}$ follows from 68.3.

14. (i) Observe that every term in the expansion of $(a - b)^{m+n-1}$ contains either a factor a^m or a factor b^n, and so belongs to the ideal J. (ii) (1) $K \neq \emptyset$ since $0 \in K$. (2) Closure of K under $-$ follows from part (i). (3) Since R commutative, $x^m \in J \Rightarrow \forall r \in R$, $(rx)^m = r^m x^m \in J$.

15. (i) C_6 is nilpotent in $\mathbb{Z}_{12}$. C_x nilpotent in $\mathbb{Z}_{pq} \Rightarrow C_{x^r}$ is zero (for some $r \in \mathbb{N}$) $\Rightarrow pq \mid x^r \Rightarrow p$ and q both divide x (cf. 12.5), etc. (ii) If $x^r = 0$, then $(x + J)^r = x^r + J = J = $ zero of R/J. (iii) If K is as described and x is arbitrary in R, then, since $x + K$ is nilpotent in R/K, $(x + K)^r = K$ for some $r \in \mathbb{N}$; hence $x^r + K = K$, i.e. $x^r \in K$; hence $(x^r)^s = 0$ for some $s \in \mathbb{N}$—which shows that x is nilpotent. (iv) If $x^n = 0$, then $(1 - x)(1 + x + x^2 + \ldots + x^{n-1}) = 1$. (v) N is an ideal by exercise 14(ii) applied with $J = \{0\}$. $(x + N)^r = N \Rightarrow x^r \in N \Rightarrow (x^r)^s = 0$ for some $s \in \mathbb{N} \Rightarrow x \in N \Rightarrow x + N = N$. (vi) It suffices to prove that an arbitrary $x \in N$ belongs to an arbitrary maximal ideal M. By part (ii), $x + M$ is nilpotent in R/M, which is a field (by 71.1). So $x + M = M$, i.e. $x \in M$.

16. $x^2 = 1 \Rightarrow (x-1)(x+1) = 0 \Rightarrow x - 1 = 0$ or $x + 1 = 0$ (D being an integral domain). *Second part*: Consider first the product of the units that are equal to neither 1 nor -1; this product is 1 because the factors in it divide into pairs of the form $\{u, u^{-1}\}$; result now follows (though the characteristic 2 case needs separate comment). *Last part*: Apply second part with $D = \mathbb{Z}_p$.

17. Suppose R/J is an integral domain. Then $J \subset R$. And $x, y \in R - J \Rightarrow x + J, y + J$ nonzero in $R/J \Rightarrow (x+J)(y+J) \neq J \Rightarrow xy + J \neq J \Rightarrow xy \in R - J$. So J prime. Converse must also be tackled. J maximal $\Rightarrow R/J$ a field $\Rightarrow R/J$ an integral domain $\Rightarrow J$ prime. Every ideal of $\mathbb{Z}$ is $\{0\}$ or (m) for some $m \in \mathbb{N}: \mathbb{Z}/\{0\} \cong \mathbb{Z}$, which is an integral domain; and $\mathbb{Z}/(m) = \mathbb{Z}_m$, which is an integral domain iff m is prime.

18. (i) Suppose that p prime in D and that $p = ab$ ($a, b \in D$). Then $p | ab$, and so $p | a$ or $p | b$—say the former (else switch notation). Then we have both $p | a$ and $a | p$; i.e. p is an associate of a. Hence one easily proves b is a unit. So the $p = ab$ factorization is trivial. Hence p irreducible. (ii) Suppose (q) is a prime ideal of D ($q \in D - \{0\}$). Since $D - (q) \neq \emptyset$, q is not a unit. Further, $q | ab \Rightarrow ab \in (q) \Rightarrow a \in (q)$ or $b \in (q)$ [since (q) is a prime ideal] $\Rightarrow q | a$ or $q | b$; and thus q is a prime element. The converse must also be tackled. Now suppose D is a principal ideal domain. (1) If J is a nonzero prime ideal of D, then (cf. (ii) above) $J = (q)$ for some prime element q; q is irreducible by part (i), and so (q) is maximal by 71.2. (2) If p is irreducible in D, then $D/(p)$ is a field (by 71.1, 71.2); hence $D/(p)$ is an integral domain; hence (by exercise 17) (p) is a prime ideal; and so (by (i) above, since $p \neq 0$) p is prime.

19. $\forall x, y \in F$, $\sigma(xy) = \sigma(x)\sigma(y)$, clearly, and $\sigma(x+y) = \sigma(x) + \sigma(y)$ because p divides the binomial coefficient $\binom{p}{r}$ if $1 \leqslant r \leqslant p - 1$. Prove σ injective by showing $\ker \sigma = \{0\}$; σ surjective follows because F is finite.

20. Suppose that θ is an isomorphism from $(F - \{0\}, .)$ to $(F, +)$. F must have characteristic 2, since otherwise $0 = \theta(1) = \theta((-1)^2) = \theta(-1) + \theta(-1) \neq 0$. If $x \in F - \{0\}$, then $\theta(x^2) = \theta(x) + \theta(x) = 0$ and hence (as θ injective) $x^2 = 1$, so that (cf. exercise 12) $x = 1$. Hence $|F| = 2$, and θ is a bijection from a set of order 1 to a set of order 2—contradiction.

21. Closure of K under subtraction: Let $x, y \in K$; then $x \in J_k$, $y \in J_l$ for some $k, l \in \mathbb{N}$, and so x and y belong to J_m, where $m = \max(k, l)$; since J_m is an ideal, $x - y \in J_m$, which is a subset of K. *Second part*: Suppose there is such a sequence of ideals, and let K be their union; K must equal (a) for some $a \in K$, and $a \in J_m$ for some $m \in \mathbb{N}$; hence $K \subseteq J_m$, and so $J_{m+1} \subseteq J_m$—contradiction.

22. Suppose R/J a field. Let K be an ideal of R with $J \subset K$. Then K contains an element x not in J. $x + J$ has an inverse $(y + J$, say) in R/J. Hence obtain $xy = j + 1$ for some $j \in J$, and use this equation to prove $1 \in K$, i.e. $K = R$, etc.

23. (i) Units must have norm 1, clearly (cf. exercise 4). (ii) $a^2 + 5b^2 = 2$, 3 have no integral solutions. (iii) Any non-trivial factorizations would involve factors of norms 2, 3. (iv) $6 = 2 \times 3 = (1 + i\sqrt{5})(1 - i\sqrt{5})$ are two inequivalent complete factorizations of 6. To prove J an ideal, use 66.4. To check condition (3), let $x + y\theta \in R$ and $a + b\theta \in J$, so that $a \equiv b \pmod 2$. Then $(x + y\theta)(a + b\theta) = (xa - 5yb) + (xb + ya)\theta$, and this belongs to J since (mod 2) $xa - 5yb \equiv xa + yb \equiv xb + ya$ (since $a \equiv b$). Note that $2 \in J$. So if $J = (d)$, $d | 2$ and hence (by parts (i) and (iii)) $d = \pm 1$ or ± 2; i.e. $J = R$ or $J = (2)$—both of which present obvious contradictions. R/K is not an integral domain since $(1 + \theta) + K$ and $(1 - \theta) + K$ are nonzero elements of R/K whose product is zero.

24. For the first part, apply 68.3 to the homomorphism $a_0 + a_1 X + \ldots + a_n X^n \mapsto a_0$. *Second part*: $D[X]$ a principal ideal domain $\Rightarrow D[X]/(X)$ a field, by 71.1, 71.2, and the fact that X is irreducible in $D[X]$.

25. Suppose the ideal J of all polynomials with even constant term equals (f). Since $2 \in J$, 2 is a multiple of f in $\mathbb{Z}[X]$. Hence $f = \pm 2$ or ± 1 —all of which possibilities obviously contradict $J = (f)$.

26. Any two generators of J are unit multiples of each other; and the units in $F[X]$ are the nonzero elements of F.

27. Let R be the ring of Gaussian integers. Let $a, d \in R$, with $d \neq 0$. If $|a| \geqslant |d|$, it is clear that, in the Argand diagram, one of the points $a + d, a - d, a + id, a - id$ is closer to the origin than a is: so one can subtract Gaussian-integer-multiples of d from a until the square of the modulus of the remainder (which can take only integral values) drops below $|d|^2$. A division algorithm result follows: there exist $q, r \in R$ such that $a = qd + r$ and $|r| < |d|$. Now let J be an arbitrary nonzero ideal of R. Choose an element d of least possible modulus in J; then use the division algorithm to prove $J = (d)$.

28. Precise statement of theorem: Let F be a field, let $f \in F[X]$, and let $\alpha \in F$. Then $(\kappa f)(\alpha) = 0$ iff $X - \alpha$ is a factor of f in $F[X]$. (Proof not difficult). If $X^2 + 1$ has a non-trivial factorization in $\mathbb{R}[X]$, then $X^2 + 1 = (X - \alpha)(X - \beta)$ for some $\alpha, \beta \in \mathbb{R}$; and above theorem yields $\alpha^2 + 1 = 0$.

Index

197